Climate Change, Politic and the Press in Ireland

Media coverage of climate change has attracted much scholarly attention because the extent of such coverage has an agenda-setting effect and because the ways in which the coverage is framed can influence public perception of and engagement with the issue. However, certain gaps in our understanding of the processes whereby such coverage is produced remain. The competition among strategic actors to influence media framing strategies is poorly understood, and the perspectives of journalists and editors are largely absent from literature. With a view to advancing our understanding of the "frame competition" around climate change and to presenting the perspectives of journalists regarding climate change as a journalistic topic, this book presents an in-depth case history of media coverage of climate change in Ireland. First, the extent of media attention for climate change is established, and the way in which such coverage is framed is also examined. Through a series of interviews, including rare and privileged access to government ministers, their media advisors, and journalists and editors, the book uncovers the contest to establish a dominant framing. The main objective of this book is to advance our understanding of the contest to establish the dominant framing of climate change in the media discourse. Although focussed on Ireland, its conclusions are of value to those seeking to better understand the dynamics of media coverage of climate change in other contexts.

This book will be of great interest to students and scholars of climate change, environmental policy, media and communication studies, and Irish politics.

David Robbins is an assistant professor in the School of Communications at Dublin City University, Ireland. He has over 25 years of experience in national print and broadcast media as a reporter, editor, and columnist.

Routledge Focus on Environment and Sustainability

The Environmental Sustainable Development Goals in Bangladesh
Edited by Samiya A. Selim, Shantanu Kumar Saha,
Rumana Sultana and Carolyn Roberts

Climate Change Discourse in Russia
Past and Present
Edited by Marianna Poberezhskaya and Teresa Ashe

The Greening of US Free Trade Agreements
From NAFTA to the Present Day
Linda J. Allen

Indigenous Sacred Natural Sites and Spiritual Governance
The Legal Case for Juristic Personhood
John Studley

Environmental Communication Among Minority Populations
Edited by Bruno Takahashi and Sonny Rosenthal

Solar Energy, Mini-grids and Sustainable Electricity Access
Practical Experiences, Lessons and Solutions from Senegal
Kirsten Ulsrud, Charles Muchunku, Debajit Palit and Gathu Kirubi

Climate Change, Politics and the Press in Ireland
David Robbins

For more information about this series, please visit: https://www.routledge.com/Routledge-Focus-on-Environment-and-Sustainability/book-series/RFES

Climate Change, Politics and the Press in Ireland

David Robbins

Routledge
Taylor & Francis Group

LONDON AND NEW YORK

earthscan
from Routledge

First published 2019
by Routledge

2 Park Square, Milton Park, Abingdon, Oxfordshire OX14 4RN
52 Vanderbilt Avenue, New York, NY 10017

Routledge is an imprint of the Taylor & Francis Group, an informa business

First issued in paperback 2020

British Library Cataloguing-in-Publication Data
A catalogue record for this book is available from the British Library

Library of Congress Cataloging-in-Publication Data
A catalog record has been requested for this book

ISBN: 978-1-138-32387-2 (hbk)
ISBN: 978-0-367-60668-8 (pbk)

Typeset in Times New Roman
by codeMantra

To my wife Fran and my daughter Grace for their love, patience, and encouragement.

Contents

Figures

Tables

Foreword

When Marshall McLuhan wrote that "All media exist to invest our lives with artificial perceptions and arbitrary values," he might have been prescient in foreseeing the crucial role of the media in presenting and interpreting the greatest challenge of the 21st century, namely climate change. Certainly, conventional science communication has failed to sensitise the general public sufficiently to the gravity of the current situation, and it has been largely through the media that their perceptions have been formed, sometimes erroneously, sometimes shaped by the hidden hand of powerful vested interests. In this text, David Robbins examines the way in which framing of the issues around climate change have occurred in Ireland. A contest for dominance is revealed, with "spin doctors," politicians, and environmental journalists vying to foreground their work by various framing strategies. The culture of each is well explored in a meticulous and forensic examination of trends over a decade involving over 700 print items and 12 semi-structured interviews with key individuals involved in decision-making and editorial judgements. This shows well the ebb and flow of climate change in the public consciousness in response to individual events and competing priorities.

David Robbins finds that Ireland's media coverage of climate change is framed predominantly around political terms, with economic, scientific, and ethical framings figuring to a lesser extent. The constraints of the news editor and journalistic culture emerge as key considerations explaining this hierarchy. Among the related issues touched on are the role of media ownership, the scientific capabilities of journalists to provide informed coverage of key scientific arguments, the roles of vested interest groups, and "celebrity" contributors in influencing the direction of travel of climate change issues in the media. To each of these, he brings his own background as a journalist turned academic and provides valuable insight. In a lucidly written text, the reader is challenged to examine his or her own position and the formative frames that may have produced it. Here, the words of McLuhan's contemporary, media scholar John Culkin, resonate: "We shape our tools and thereafter our tools shape us."

John Sweeney,
Emeritus Professor of Geography,
Maynooth University
Contributing author to IPCC AR4

so lets fashion new tools

- I would go further to say - we need an *urgent* message

1 Introduction

Introduction

In December 2015, Enda Kenny, then Taoiseach (prime minister) of Ireland, flew to France to address the 21st Conference of the Parties to the UN Framework Convention on Climate Change at Le Bourget, on the outskirts of Paris. He was one of over 80 world leaders to do so (Agence France Presse, 2015), and his address formed part of the elaborate choreography of the climate change talks. In Copenhagen in 2009 – the last time a binding global agreement on climate change seemed likely – the leaders arrived at the end of the talks to confer their blessing on what their negotiating teams had agreed. Things did not go as planned; the late arrival of heads of government was counterproductive, and only a modest agreement – the Copenhagen Accord – was signed (Bodansky, 2010). In Paris, world leaders arrived at the outset of the talks instead. Each gave a short, encouraging address, intended to set a positive tone for the talks, and departed. Enda Kenny's speech was typical: he expressed hope that a binding global agreement limiting carbon emissions would be signed. In a key passage, he stated,

> I hope that we are serious about putting in place a legally binding agreement on climate change that will underpin our actions on the goals already agreed and enhance our ability to reach them. This requires action by everybody – big and small. Ireland is determined to play its part.
>
> (*Irish Times*, 2015; Kenny, 2015)

However, Kenny also gave a separate briefing to the Irish media covering the event and stated that in fact, tackling climate change was "not a priority" for his government (McGee, 2015; McGee and Marlowe, 2015; Melia, 2015; Robbins, 2015). These events provide a useful lens through which to examine the ways the arenas of Irish politics, policy, and media interact when it comes to climate change. There is the articulation of support for the high ideals of international cooperation and climate action, echoing Kenny's remarks to the General Assembly of the United Nations on the UN's Sustainable Development Goals not long before (Kelly, 2015), undercut by a more pragmatic approach which seeks to protect perceived national interests, especially agriculture (An Taisce, 2015; Gibbons, 2015). It is this nexus, where media, politics and policy meet, that forms the primary focus of this work.

Why is media coverage of climate change important?

Climate change is happening at a physical level on our planet. Scientists study its effects, bureaucracies develop policies to respond, and politicians debate whether or how to implement these measures. Does it matter how often, or in what manner, the media covers the issue? I would argue that it does. First, the media has an agenda-setting effect; the more media coverage an event or an issue receives, the more important the public thinks it is (McCombs and Shaw, 1972; Ungar, 1992; McCombs, 2004). Public amplification in the media "provides a certification of importance" (Schudson, 1995, p. 20); elsewhere, the same author adds that "When the news media offer the public an item of news, they confer on it a public legitimacy." (Schudson, 2011, p. 22). The media are "central agents for *raising awareness* and *disseminating information* (Schäfer et al., 2014, p. 1233, emphasis in original), and their coverage of climate change may create circumstances "where it is conducive for governments to act, or hard for them not to act in the face of perceived pressure to initiate a policy response." (Newell, 2000, p. 94). Furthermore, mass media are an important forum in which various responses to climate change are discussed and legitimated (Nanz and Steffek, 2004, p. 321; Schneider, Nullmeier and Hurrelmann, 2007, p. 136). They are also a forum in which various societal actors – environmental non-governmental organisations, business associations, political parties, policymakers – may put across their views and, in turn, have these views exposed to feedback from others (Steffek, 2009, p. 315). Where there is disagreement about the effects of climate change, or even the extent to which is it anthropogenic, or about mitigation or adaptation measures which may or may not be undertaken, the media helps clarify the positions of the various societal actors on the issue (Peters, 2008, p. 131).

The media also perform the function of informing the public about matters that may be remote from them physically or otherwise unobtrusive. Many people in Western Europe, for example, do not as yet witness the effects of climate change in their daily lives. They do not experience the retreat of glaciers or the rise of sea levels or the acidification of the oceans; they rely on the media to inform them about such matters. The public gets most of its information about science from the mass media (Nelkin, 1987; Wilson, 1995). There is a further reason why the extent and nature of media coverage of climate change is important: aspects of media coverage have an effect on how the general public thinks about the issue. The particular features of climate change which are accentuated by the media shape how people respond to it. If it is framed as an economic issue, or as a moral issue, or as purely scientific issue, that is how the consumers of media tend to view it. The media is important in societal responses to climate change for these four reasons: the media educates the public about climate change; the media alerts the public to its relative importance; they suggest how people should think about it; and they provide a space in which the positions of various parties can be articulated, clarified, or contested.

Climate change as a "wicked" problem

The Fifth Assessment Report (AR5) of the United Nations Intergovernmental Panel on Climate Change (IPCC), published in November 2014, makes it clear that "Continued emission of greenhouse gases will cause further warming and long-lasting changes in all components of the climate system, increasing the likelihood of severe, pervasive and irreversible impacts for people and ecosystems" (IPCC, 2014, p. 2). The series of reports that comprise AR5 emphasise that climate change is a serious threat requiring international action. At the time of writing, AR5 is the latest in a series of assessments of the effect that greenhouse gases, or GHGs (predominantly carbon dioxide [CO_2], methane, and nitrous oxide, although chlorofluorocarbons and sulphur hexafluoride contribute to the greenhouse effect also) are having on the atmosphere. The first of these reports was published in 1990 and the others have followed every five, six, or (as in the case of AR5) seven years. The language concerning the anthropogenic contribution to climate change in the reports has become more definitive over time, with AR5 declaring that "warming of the climate system is unequivocal" (IPCC, 2014, SPM 1.1, p. 2) and that

Anthropogenic greenhouse gas emissions have increased since the pre-industrial era, driven largely by economic and population growth, and are now higher than ever. This has led to atmospheric concentrations of carbon dioxide, methane and nitrous oxide that are unprecedented in at least the last 800,000 years. Their effects, together with those of other anthropogenic drivers, have been detected throughout the climate system and are *extremely likely* to have been the dominant cause of the observed warming since the mid-20th century.

(Ibid., SPM 1.2, p. 4, emphasis in original)

The Assessment Reports of the IPCC are representative of a broad scientific consensus on the anthropogenic contribution to climate change. The National Academies in the United States have also confirmed the human influence in rising levels of CO_2 and other GHGs (Brand, 2010, p. 15). A study of 11,944 scientific papers on climate change published between 1991 and 2011 showed that between 97.1% and 97.2% of scientific literature which expressed a position on the origins of climate change endorsed the existence of anthropogenic global warming and described the number of papers rejecting the consensus on anthropogenic global warming as a "vanishingly small proportion of published research" (Cook et al., 2013, p. 1). This has led some to describe the current geological period, previously known as the Holocene, as the "Anthropocene," due to the increasing impact of humans on the planet (Crutzen, 2006; Steffen, Crutzen and McNeill, 2007).

Ireland has contributed to, and has been affected by, global warming. During the period 1995–2005, the country witnessed changes in settlement patterns, deteriorating environmental quality, and considerable expansion of transport, waste, and water usage (Pape et al., 2011, p. 29). Indeed, Ireland recorded "the highest percentage increase of greenhouse gas emissions from the transport sector of any EU state during the period 1990–2003" (Flynn, 2007, p. 58). Provisional measurements show that Ireland's total GHG emissions for 2016 amounted to 61.19 million tonnes of CO_2 equivalent (Mt CO_2 eq.; EPA Ireland, 2017), and Ireland is one of only 12 parties to the United Nations Framework Convention on Climate Change (UNFCCC) to show an increase in total aggregate anthropogenic emissions of GHGs between 1990 and 2015 (UNFCCC, 2017). Ireland's per capita emissions are relatively high; the country is the third-highest per capita emitter in the EU (Ó Fátharta, 2016). This elevated per capita figure is largely due to the distorting effect of emissions from the agriculture sector, which are expected to account for over 45% of Ireland's non-ETS emissions by 2020. Ireland is required by the EU to reduce its non-ETS emissions by 20% (as measured against 2005 levels) by 2020 (Environmental

Protection Agency, 2014). It is expected to miss the 2020 target by a wide margin (Climate Change Advisory Council, 2017; Melia, 2017). In addition, Ireland is, at the time of writing, involved in EU-wide negotiations regarding an Effort Sharing Regulation which will set emissions targets for 2030 (Environmental Protection Agency, 2014). Ireland has been criticised for adopting a negotiating position aimed at reducing the ambition of the 2030 targets (O'Sullivan, 2017b).

Meanwhile, the impacts of climate change are beginning to be felt in Ireland, with increased rainfall, fewer frost nights, and more frequent flood events (Sweeney, 2000; McElwain and Sweeney, 2007; Environmental Protection Agency, 2009), with implications for Ireland's agriculture (Sweeney et al., 2003, 2007), fisheries (Pinnegar, Buckley and Englehard, 2017), and economy (Flood, 2012; Government of Ireland, 2017).

Since the publication of AR5, the findings of climate science suggest that the models used by the IPCC have underestimated the rate of warming and the extent of impacts (Harrison et al., 2016; Stern, 2016). However, political responses to climate change have been largely inadequate (Giddens, 2009), and it is considered that even the international agreement to reduce global emissions arrived at in Paris in December 2015 may not include sufficient measures to keep global warming to below 2C (Anderson and Peters, 2016a, 2016b).

Why the Irish experience is relevant

As Torney and Little point out, "Ireland is not just an understudied case. In many respects, it is an interesting case, both for its peculiarities and for what it shares with other countries" (2017, p. 193). In terms of transnational politics and EU governance, Ireland is a small, peripheral country and has traditionally been a "rule-taker" as far as the EU is concerned, rather than a "rule-maker" (Van de Graff, 2017). In this respect, Ireland has much in common with other small EU states, such as Denmark, Finland, Estonia, Slovakia, and Slovenia, and with small island members of the EU, such as Malta and Cyprus. Rule-taker states typically rely on a larger administrative bloc to provide regulatory frameworks and other policy initiatives (Björkdahl, 2008) and often have difficulty in influencing central policy in such blocs, although there have been some exceptions (Kronsell, 2018). Ireland's small size and relatively low population density may make it easier to sustain concerted action, as small size is associated with agility at government level and the capacity for consensual policy development (Katzenstein, 1985; Keating, 2015). Furthermore, the relative lack of ideological polarisation within Ireland's party political system may also mean that policies enacted by one party are unlikely to be dispensed with by succeeding governments. More recently, Ireland has aligned itself with Poland and Lithuania in lobbying to reduce the ambition of EU emissions reduction targets (O'Sullivan, 2017b). In terms of its role and influence within the EU, Ireland has much in common with the Nordic states and other small countries, many of whom have been at the forefront of environmental policy development at the EU and national levels (Liefferink and Andersen, 1998), while also forming coalitions with larger and less environmentally focused countries. When it comes to climate change legislation and other policy instruments to help tackle climate change, at first glance, Ireland seems to have adopted a forward-looking and progressive approach. It has implemented a carbon tax, for instance, and has encouraged renewable energy development. It has also developed an agenda of emissions reduction in agriculture, through its policy of "climate smart agriculture" (Curtin and Arnold, 2016), and has joined the ranks of countries to have enacted framework climate legislation (Government of Ireland, 2015). Ireland also established a Citizens' Assembly, an exercise in deliberative democracy, which discussed climate change and issued a report calling for the withdrawal of subsidies for peat and coal energy production (The Citizens' Assembly, 2018). Other initiatives are discussed in more detail later, but it is important to note for now that Ireland appears to have taken action at the government level on climate change. However, on closer inspection, many of these measures and initiatives take the form of calls and imprecations, of reports outlining the challenges of dealing with Ireland's emissions rather than policies to actually reduce them.

Ireland is also a resonant case due to the impact on its society and economy of the global financial crisis of 2008. The circumstances as they relate to Ireland, and the impact on Ireland's willingness to engage with climate change as an issue, are discussed elsewhere. However, it is evident that, alongside countries such as Spain, Portugal, and Greece (the so-called "PIGS" countries), the crisis had a considerable impact on every aspect of Ireland's society. The response of the Irish media to the crisis, and the impact of the crisis on the media discourse of climate change, explored here, may hold lessons for other territories.

The presence of the Green Party in government in Ireland for the first time, and the impact of this development on the level and nature of media attention for climate change, may also be of interest to students of press/politics. The ways in which Green Party politicians communicated about climate change, the ways in which they attempted to influence the media's framing of the issue, and the response of journalists and editors to such strategies are relevant to other territories such as Germany, Denmark, Finland, Belgium, France, and Italy, where Green Parties have served in coalition governments.

Ireland's media system is also a point of commonality – and of difference – with other territories. According to Hallin and Mancini, Ireland belongs to the North Atlantic or Liberal model, reflecting a medium newspaper circulation, neutral commercial press, and strong professionalisation within the journalism sector. In the national discourse around political and commercial orientation, the dichotomy of "Boston or Berlin" is often cited as a key question. In terms of media orientation and tradition, Ireland is more in the UK/US tradition than the European tradition, retaining, as it does, an emphasis on the separation of reporting and comment rather than the more discursive and polemical journalism of Europe (Chalaby, 1996; Schudson, 2001). However, Ireland does not exhibit the polarisation and partisan coverage of climate change evident in the US, the UK, and Australia (Painter and Ashe, 2012); its coverage of the issue has more in common with that evident in France and Germany. In some ways, Ireland is at the intersection of two differing media systems: the US/UK system, with its mixture of rowdy tabloids and high-minded broadsheets, and the European system, with its emphasis on discourse and its quasi-literary status. How Ireland's media system responds to climate change may further our understanding of both of these media systems.

Lastly, Ireland, along with countries such as Denmark and New Zealand, relies heavily on its agriculture and food sector. Agriculture contributes 33% of Ireland's total emissions, compared to an EU average of 10%. Only New Zealand has a higher proportionate level of agricultural emissions, at 49% (New Zealand Agricultural Greenhouse Gas Research Centre, 2016). The Irish agri-food sector accounts for 7.6% of GDP and 8.6% of total employment (Department of Agriculture and Food, 2017). Given the commercial and cultural influence of the agriculture industry in Ireland, how does the news media treat this sector's role in contributing to climate change? I believe the case of Ireland's media and their reporting of agriculture and climate change is of interest to scholars and policymakers in other territories where there is a substantial contribution of ruminant animals and agricultural land husbandry to national emissions profiles. In summary, Ireland's position as a small EU state on the periphery of the bloc, its experiences during the global financial crisis of 2008, the presence of the Green Party in government, its agricultural emissions profile, and its bridging position between the US and the UK, make it a compelling, and neglected, focus of study.

The policy, political, and media context of climate change in Ireland

The policy response to climate change in Ireland has been slow, leading to the suggestion that Ireland is a "laggard" when it comes to climate policy (Torney and Little, 2017). More recently, in a question and answer session following his speech to the European Parliament in January 2018, Ireland's Taoiseach admitted he was "not proud" of Ireland's record in tackling climate change and that Ireland was "a laggard" on the issue (Sargeant, 2018). On September 30 and October 1, 2017, Ireland's Citizens' Assembly (in which 100 citizens consider social issues) considered the topic of "How the State can make Ireland a leader in tackling climate change." The Citizens' Assembly meeting was the most recent in a range of climate-related publications and initiatives, including a National Mitigation Plan (Government of Ireland, 2017) and a public consultation process entitled the National Dialogue on Climate Action, mandated by the Programme for Government agreement between Fine Gael and several independent deputies. Some of these initiatives were mandated in Ireland's primary climate change legislation, the Climate Action and Low Carbon Development Act (Government of Ireland, 2015). In general, the policy responses to climate change in Ireland, including the Act, have been roundly criticised for (i) the absence of binding sectoral targets for emissions reduction and (ii) the lack of specific policy proposals as to how a transition to a low-carbon economy and society is to be achieved (Torney, 2017). Indeed, the Climate Change Advisory Council, an expert body established by the 2015 Act, has warned that Ireland's emissions are in fact rising in the agriculture and transport sectors, and that "in the absence of further policies or measures, the overshoot on emissions could be even larger than currently projected" (Climate Change Advisory Council, 2017, p. 10). At the meeting of the Citizens' Assembly, it was remarked that making Ireland a leader in tackling climate change was "fanciful," that Ireland was failing "spectacularly" to meet its obligations on emissions reduction, and that it might be a better first step to ensure Ireland simply did its fair share (O'Sullivan, 2017a).

Political progress in advancing climate legislation has also been a slow and contested process, with sectoral interests, particularly the agriculture lobby, arguing against specified targets. At various points since Ireland was assigned emissions targets as a result of the burden-sharing agreement of the Kyoto Protocol in 1998, a range of political parties, including the Green Party, Labour, and Fine Gael, have been involved in proposing climate legislation (Torney, 2017).

A comprehensive bill was proposed by the Labour Party in 2007 and another by the Green Party in 2010; neither was enacted. A less stringent bill was proposed by a Fine Gael minister in 2013, and a version of this legislation was again presented by his Labour Party successor in 2015, at which point it was passed by the Oireachtas (Ibid.).

The Labour Party bill of 2007 was introduced as a private member's bill in the Seanad (upper house) when the Labour Party was in opposition and stood little chance of being accepted by the government. The 2010 bill, put forward by the Green-Fianna Fail government, had been accepted at cabinet and was in a consultation period when that government fell. The 2015 Act was passed by a Fine Gael-Labour coalition. An interesting difference in the drafting of the 2010 Green-led legislation and the 2015 Fine Gael-led Act relates to the composition of a "national expert advisory body on climate change" (now called the Climate Change Advisory Council, mentioned earlier). In the 2010 bill, the chief executives of the Environmental Protection Agency and the Sustainable Energy Association of Ireland were proposed as ex officio members of this body; in the 2015 legislation, these were augmented by the director of Teagasc, the state agricultural research agency, and the director of the Economic and Social Research Institute, signalling a turn away from an environmental emphasis and towards an agricultural and economic one.

The recent emergence of a climate contrarian organisation in Ireland, entitled the Irish Climate Science Forum, has challenged a broad social and political consensus on the anthropogenic origins of recent climate change (Gibbons, 2017). The group has, at the time of writing, hosted two closed meetings at which noted climate deniers (Richard Lindzen and William Happer) addressed the audience. Lindzen was the subject of an open letter from Massachusetts Institute of Technology scientists dissociating themselves from Lindzen's appeal to President Trump urging the US to withdraw from the Paris climate change agreement (Hirji, 2017), while Happer, among those being considered for the post of science advisor to the White House, has espoused a range of denier positions (Readfearn, 2017). Both Happer and Lindzen questioned the contribution of methane to global warming. The meetings were covered by the media (even though journalists were prevented from attending), and the claim that methane was contributing less to the warming effect than previously thought was dismissed as "balderdash" by eminent climate scientist John Sweeney (Boucher-Hayes, 2017).

Ireland, then, is a small country with a large per capita emissions profile, whose emissions from the transport and agriculture sectors are rising; Ireland will not meet its emissions reductions targets in 2020 and has not produced any policy initiatives that might help it to meet its 2030 targets. Despite some recent climate-related initiatives, Ireland remains a laggard in terms of climate legislation, and the issue remains low on the political agenda.

It is helpful to sketch the media landscape as it pertains to the newspaper market. Sales of print newspapers are dominated by Independent News and Media, which publishes several leading titles, including the *Irish Independent*, the *Sunday Independent*, the *Sunday World*, and the *Evening Herald*. The Irish Times Trust Ltd. publishes the *Irish Times* and has just completed the purchase of the *Irish Examiner* from Landmark Media. Several UK media organisations publish Irish editions, such as News International (the *Times Ireland* and the *Sunday Times* Ireland edition, the *Irish Sun*), Trinity Mirror (the *Irish Mirror*), and Associated Newspapers (the *Irish Daily Mail*). Some 76% of the Irish population read a newspaper between two and three times per week (Flynn, 2016), although newspaper circulation and readership is falling. The circulation and readership figures for the newspapers studied in this book are given in Table 1.1.

Although Ireland records a low score on the Media Pluralism Monitor, the score for concentration of media ownership is high, principally due to the influence of businessman Denis O'Brien, who has extensive cross-media holdings in radio, online, and print media and is the largest shareholder in Independent News and Media (Flynn, 2016). The media market in Ireland is quite diffuse, with legacy titles such as those of the Independent group and the *Irish Times* continuing to be influential, while net native news outlets such as

Table 1.1 Circulation and readership of seven national Irish newspapers

Title	Number of copies sold	Readership
Irish Independent	95,502	508,000
Irish Times	62,423	317,000
Irish Examiner	28,338	194,000
Irish Daily Mail	41,027	172,000
Sunday Independent	185,080	852,000
Sunday Business Post	30,202	109,000
Sunday Tribune	54,400	171,000

Source: Newsbrands Ireland.

Note: the *Sunday Tribune* ceased publication in 2011. The *Tribune*'s circulation and readership figures relate to 2010.

Circulation figures relate to the period from January to June 2017; readership figures relate to 2014–2015.

thejournal.ie grow their audiences. However, the state, in the form of Radio Telefís Éireann's television, radio, and online news offerings, dominates the media landscape (McNamara et al., 2017). This brief outline is intended to give a sense of the media environment in Ireland and to establish the reach and the agenda-setting potential of the titles examined. The six titles listed earlier have a combined circulation of 496,972 copies and a combined readership of 2,323,000, representing considerable market penetration. Critical perspectives on the Irish media landscape from journalism history (O'Brien, 2017), political economy (McCullagh, 2008; Cawley, 2012; Preston and Silke, 2014), journalism practice (Heravi, Harrower and Boran, 2014), and more general perspectives on the role of the media in Irish society (Bell, 1986; Horgan, 2001; Truetzschler, 2007; Horgan and Flynn, 2017) are acknowledged; however, the critical focus of this work remains the framing strategies of those communications about climate change during the timeframe examined.

References

Agence France Presse (2015) 'World leaders to attend Paris climate summit', *The Guardian*, 28 October.

An Taisce (2015) 'Worst possible start to COP21 for Ireland as Taoiseach drops ball on Day 1', *Press statement released by An Taisce*, November 30. Available at www.antaisce.org/articles/worst-possible-start-to-cop21-for-ireland-as-taoiseach-drops-ball-on-day-1

Anderson, K. and Peters, G. (2016a) 'Act now, not tomorrow', *New Scientist*, pp. 19–20.

Anderson, K. and Peters, G. (2016b) 'The trouble with negative emissions', *Science*, 354(6309), pp. 182–183.

Bell, D. (1986) *Is the Irish Press Independent? Essays on Ownership and Control of the Provincial, National and International Press in Ireland*. Dublin: Media Association of Ireland.

Björkdahl, A. (2008) 'Norm advocacy: A small state strategy to influence the EU', *Journal of European Public Policy*, 15(1), pp. 135–154.

Bodansky, D. (2010) 'The Copenhagen climate change conference : A postmortem', *American Journal of International Law*, 104(2), pp. 230–240.

Boucher-Hayes, P. (2017) 'Irish climate science forum', *Drivetime*, RTE Radio 1, broadcast May 10, 2018. Podcast available at www.mixcloud.com/rt%C3%A9radioplayerlatestpodcasts/drivetime-irish-climate-science-forum/

Brand, K.-W. (2010) 'Social practices and sustainable consumption: Benefits and limitations of a new theoretical approach', in Gross, M. and Heinrich, H. (eds.) *Environmental Sociology: European Perspective and Interdisciplinary Challenges*. Netherlands: Springer, pp. 217–235.

Cawley, A. (2012) 'Sharing the pain or shouldering the burden?', *Journalism Studies*, 13(4), pp. 600–615.

Chalaby, J. (1996) 'Journalism as an Anglo-American invention. A comparison of the development of French and Anglo-American journalism', *European Journal of Communication*, 3, pp. 303–326.

Climate Change Advisory Council (2017) *Periodic Review Report 2017*. Available at http://www.climatecouncil.ie/media/CCAC_PERIODICREVIEW REPORT2017_Final.pdf

Cook, J., Nuccitelli, D., Green, S.A., Richardson, M., Winkler, B., Painting, R., Way, R., Jacobs, P. and Skuce, A. (2013) Quantifying the consensus on anthropogenic global warming in the scientific literature. *Environmental Research Letters*, 8(2), p. 024024.

Crutzen, P.J. (2006) 'The "anthropocene"', in *Earth System Science in the Anthropocene*. Berlin; Heidelberg: Springer, pp. 13–18.

Curtin, J. and Arnold, T. (2016) *A Climate-Smart Pathway for Irish Agricultural Development. Exploring the Leadership Opportunity*. The Institute for International and European Affairs, Dublin.

Department of Agriculture and Food (2017) *Irish Agri Food Sector – May 2017*.

Environmental Protection Agency, Ireland (2014) *Ireland's Greenhouse Gas Emission Projections, 2013–2030*. Report. Available at www.epa.ie/pubs/reports/air/airemissions/GHG_report2014.pdf

Environmental Protection Agency Ireland (2009) *A Summary of the State of Knowledge on Climate Impacts for Ireland*, Climate Change Research Programme (CCRP) 2007–2013. Report. Available at https://www.epa.ie/pubs/reports/research/climate/CCRP1(low).pdf

Environmental Protection Agency Ireland (2017) *Ireland's Provisional Greenhouse Gas Emissions in 2016*. Report. Available at http://www.epa.ie/pubs/reports/air/airemissions/ghgemissions2016/Report_GHG%201990-2016 %20April_for%20Website-v3.pdf.

Flood, S. (2012) 'Climate change and potential economic impacts in Ireland: The case for adaptation', *Unpublished thesis*. Department of Geography, Maynooth University, p. 247.

Flynn, R. (2016) *Media Pluralism Monitor 2016: Ireland*.

Gibbons, J. (2015) 'Never mind the bullocks, Enda isn't cowed at COP', *Think or Swim Blog*. Available at www.thinkorswim.ie/2657-2/

Gibbons, J. (2017) *New Climate Science Denial Group Launches in Ireland, Desmog UK*. Report. Available at www.desmog.co.uk/2017/05/05/new-climate-science-denial-group-launches-ireland

Giddens, A. (2009) *The Politics of Climate Change*. Malden, MA: Polity Press.

Government of Ireland (2015) *Climate Action and Low Carbon Development Act 2015*.

Government of Ireland (2017) *National Mitigation Plan*.

Van de Graff, T. (2017) 'Rule-maker of rule-taker? The EU and the shifting global political economy of energy', in Leal-Arcas, R. and Wouters, J. (eds.) *Research Handbook on EU Energy Law and Policy*. Cheltenham, UK and Northampton, MA: Edward Elgar Publishing, pp. 165–178.

Harrison, P.A. et al. (2016) 'Climate change impact modelling needs to include cross-sectoral interactions', *Nature Climate Change*, 6(9), pp. 885–890.

Heravi, B.R., Harrower, N. and Boran, M. (2014) 'Social journalism survey: First national survey on Irish journalists' use of social media', *Report. The Digital Humanities and Journalism Group, Insight Centre for Data Analytics, National University of Ireland, Galway.* Galway, pp. 1–32.

Hirji, Z. (2017) 'Climate contrarian gets fact-checked by MIT colleagues in open letter to Trump', *Inside Climate News*, 6 March.

Horgan, J. (2001) *Irish Media: A Critical History Since 1922.* London: Routledge.

Horgan, J. and Flynn, R. (2017) *Irish Media: A Critical History.* Dublin: Four Courts Press.

IPCC (2014) 'Climate change 2014 synthesis report summary chapter for policymakers', *Ipcc.*

Ireland, G. (2017) *Draft National Mitigation Plan.*

Irish Times (2015) 'COP21: Full text of Taoiseach Enda Kenny's address to Paris summit', 30 November.

Katzenstein, P.J. (1985) *Small States in World Markets: Industrial Policy in Europe.* Ithaca, NY: Cornell University Press.

Keating, M. (2015) 'The political economy of small states in Europe', in Baldersheim, H. and Keating, M. (eds.) *Small States in the Modern World: Vulnerabilities and Opportunities.* Cheltenham, UK and Northampton, MA: Edward Elgar Publishing, pp. 1–19.

Kelly, F. (2015) 'Action on ending poverty and hunger at core of Enda Kenny's UN address', *Irish Times*, 25 September.

Kenny, E. (2015) 'National statement delivered by An Taoiseach Enda Kenny to the 21st conference of the parties to the UN framework convention on climate change', *Speech.* Available at www.taoiseach.gov.ie/eng/News/Taoiseach%27s_Speeches/National_Statement_by_An_Taoiseach_Enda_Kenny_T_D_21st_Conference_of_the_Parties_to_the_UN_Framework_Convention_on_Climate_Change_30_November_2015.html

Kronsell, A. (2018) 'Can small states influence EU norms?: Insights from Sweden's participation in the field of environmental politics', *Scandinavian Studies*, 74(3), pp. 287–304.

Liefferink, D. and Andersen, M.S. (1998) 'Strategies of the "green" member states in EU environmental policy-making', *Journal of European Public Policy*, 5(2), pp. 254–270.

McCombs, M.E. (2004) *Setting the Agenda: The Mass Media and Public Opinion.* Cambridge: Polity Press.

McCombs, M.E. and Shaw, D.L. (1972) 'The agenda-setting function of mass media', *The Public Opinion Quarterly*, 36(2), pp. 176–187.

McCullagh, C. (2008) 'Modern Ireland, modern media, same old story?', in O'Sullivan, S. (ed.) *Contemporary Ireland: A Sociological Map.* Dublin: UCD Press, pp. 136–151.

McElwain, L. and Sweeney, J. (2007) *Key Meteorological Indicators of Climate Change in Ireland.*

McGee, H. (2015) 'Paris climate talks: Irish have sent out mixed messages', *Irish Times*, 10 December.

McGee, H. and Marlowe, L. (2015) 'COP21: Kenny criticises "unrealistic" climate targets', *Irish Times*, 1 December.

McNamara, P. et al. (2017) *Reuters Digital News Report 2017: Ireland.*

Melia, P. (2015) 'Climate change is not our priority – Taoiseach', *Irish Independent*, 1 December.

Melia, P. (2017) 'Ireland set to miss its 2020 emissions target by "a substantial margin"', *Irish Independent*, 26 July.

Nanz, P. and Steffek, J. (2004) 'Global governance, participation and the public sphere', *Government and Opposition: An International Journal of Comparative Politics*, 39(2), pp. 314–355.

Nelkin, D. (1987) 'Selling science: How the press covers science and technology', *Physics Today*, 43, pp. 41–49.

New Zealand Agricultural Greenhouse Gas Research Centre (2016) *Agricultural Greenhouse Gases: What We Are Doing.*

Newell, P. (2000) *Climate for Change: Non-State Actors and the Global Politics of the Greenhouse Effect.* Cambridge: Cambridge University Press.

O'Brien, M. (2017) *The Fourth Estate: Journalism in Twentieth-Century Ireland.* Manchester: Manchester University Press.

O'Sullivan, K. (2017a) 'Ireland failing "spectacularly" in response to climate change', *Irish Times*, 30 September.

O'Sullivan, K. (2017b) 'Ireland taking a "shameful" approach to emissions targets', *Irish Times*, 13 October.

Ó Fátharta, C. (2016) 'Ireland's CO_2 emissions third highest in EU', *Irish Examiner*, 23 November.

Painter, J. and Ashe, T. (2012) 'Cross-national comparison of the presence of climate scepticism in the print media in six countries, 2007–10', *Environmental Research Letters*, 7(4), p. 044005.

Pape, J., Rau, H., Fahy, F. and Davies, A., 2011. Developing policies and instruments for sustainable household consumption: Irish experiences and futures. *Journal of Consumer Policy*, 34(1), pp.25–42.

Peters, B. (2008) 'The functional capacity of contemporary public spheres', in Wessler, H. (ed.) *Public Deliberation and Public Culture*. Basingstoke: Palgrave Macmillan, pp. 121–133.

Pinnegar, J.K., Buckley, P. and Englehard, G. (2017) 'Impacts of climate change in the UK and Ireland', in Phillips, B.F. and Perez-Ramirez, M. (eds.) *Climate Change Impacts on Fisheries and Aquaculture: A Global Analysis.* New York: John Wiley & Sons.

Preston, P. and Silke, H. (2014) 'Ireland—from Neoliberal champion to "the Eye of the Storm"', *Javnost – The Public*, 21(4), pp. 5–23.

Readfearn, G. (2017) 'Trump's potential science adviser William Happer: Hanging around with conspiracy theorists', *The Guardian*, 21 February.

Robbins, D. (2015) 'The UN meets MoJo: How journalists covered the Paris COP', *Blog Post*. Institute for Future Media and Journalism, Dublin City University.

Sargeant, N. (2018) *Taoiseach tells EU he is not proud of Ireland's role as Europe's climate 'laggard'*, Green News.ie.

Schäfer, M.S. et al. (2014) 'What drives media attention for climate change? Explaining issue attention in Australian, German and Indian print media from 1996 to 2010', *International Communication Gazette*, 76(2), pp. 152–176.

Schneider, S., Nullmeier, F. and Hurrelmann, A. (2007) 'Exploring the communicative dimension of legitimacy: Text analytical approaches', in Hurrelmann, A., Schneider, S. and Steffek, J. (eds.) *Legitimacy in an Age of Global Politics*. London: Palgrave Macmillan UK, pp. 126–155.

Schudson, M. (1995) *The Power of News*. Cambridge, MA: Harvard University Press.

Schudson, M. (2001) 'The objectivity norm in American journalism', *Journalism*, 2(2), pp. 149–170.

Schudson, M. (2011) *The Sociology of News*. 2nd edn. New York; London: W. W. Norton.

Steffek, J. (2009) 'Discursive legitimation in environmental governance', *Forest Policy and Economics*, 11(5–6), pp. 313–318.

Steffen, W., Crutzen, P.J. and McNeill, J.R. (2007) 'The anthropocene: Are humans now overwhelming the great forces of nature', *AMBIO: A Journal of the Human Environment*, 36(8), pp. 614–621.

Stern, N. (2016) 'Economics: Current climate models are grossly misleading', *Nature*, 530(7591), pp. 407–409.

Sweeney, J. (2000) 'A three-century storm climatology for Dublin 1715–2000', *Irish Geography*, 33(1), pp. 1–14.

Sweeney, J. et al. (2003) *Climate Change: Scenarios & Impacts for Ireland. Final Report*.

Sweeney, J. et al. (2007) *Climate Change – Refining the Impacts for Ireland*.

The Citizens' Assembly (2018) *Third Report and Recommendations of the Citizens' Assembly*.

Torney, D. (2017) 'If at first you don't succeed: The development of climate change legislation in Ireland', *Irish Political Studies*, 32(2), pp. 247–267.

Torney, D. and Little, C. (2017) 'Symposium on the politics of climate change in Ireland', *Irish Political Studies*, 32(2), pp. 191–198.

Truetzschler, W. (2007) 'The Irish media landscape', in Terzis, G. (ed.) *European Media Governance: National and Regional Dimensions*. Bristol: Intellect Books, p. 33.

UNFCCC (2017) *National Greenhouse Gas Inventory Data for the Period 1990–2015, Framework Convention on Climate Change*.

Ungar, S. (1992) 'The rise and (relative) decline of global warming as a social problem', *The Sociological Quarterly*. Wiley Online Library, 33(4), pp. 483–501.

Wilson, K.M. (1995) 'Mass media as sources of global warming knowledge', *Mass Communications Review*, 22(1–2), pp. 75–89.

2 Climate change as a complex social problem

Introduction

Those interested in how often – and how – the media cover climate change are often motivated by a concern over the disparity between the urgency of the problem of climate change and the lack of urgency in political and policy responses to do anything about it. The media's role in this state of affairs, given the agenda-setting power of the news media, is naturally of interest. However, there is a broader theoretical approach to issues such as climate change – complex social problems requiring responses across a broad range of social arenas – which is concerned with how and why certain issues rise to the top of the policy agenda while others, equally deserving of attention, languish at the bottom. These theories of agenda systems consider the realms of the social, political, and policy, and while the media's role is acknowledged in some of these agenda models, they are not given the prominence or power of agency accorded them by media scholarship. In considering how the media cover climate change, it is necessary, first of all, to situate the media in these broader agenda systems.

How issues rise and fall: theories of agenda systems

There is no shortage of research into how issues "spring to life" (Djerf-Pierre, 2012b, p. 499), but less on how they "fall from grace," on "why real problems fall from the news – beyond the cliché that they become 'stale'" (Mazur, 1998, p. 470). A paper published by Anthony Downs in 1972 attempted an explanation: *issue attention cycles*. Issues go through a predictable five-stage cycle, he argued:

i   A pre-problem stage, in which "some highly undesirable condition exists" but has yet to come to public attention;
ii An alarmed discovery and euphoric enthusiasm stage, in which a series of dramatic events (Downs offers the example of ghetto

riots) both informs and alarms the public. There is great optimism about society's ability to "solve" the problem in this stage also;

iii Realising the cost of significant progress, a stage in which it comes to be understood that "solving" the problem is going to be expensive and will involve sacrifice by a large section of society;

iv Gradual decline of intense public interest;

v A post-problem stage, in which the issue retreats to a "twilight world," but retains the ability to sporadically recapture national interest.

(Downs, 1972, pp. 39–41)

Downs set forth his model "in largely anecdotal fashion" (Howlett, 1997, p. 7), and his criteria for the types of issues that go through the issue cycle process left many important social problems out. In a study of a decade of environmental coverage in the US, Downs's model was found to be a "good fit" for the pattern of coverage and "can be used as a more general basis" for examining and dividing patterns of coverage (Trumbo, 1996, p. 280). Other researchers have applied the model to media treatment of climate change (Brossard, Shanahan and McComas, 2004; Nisbet and Huge, 2006), but it has been criticised for being vague (Howlett, 1997), too inflexible, too linear, and in focusing on the media and public agendas, it does not account for other influences (Anderson, 2009, p. 169).

In their **punctuated equilibrium** theory of how changes in policy come about, Baumgartner and Jones (1993) use evolutionary theory to argue that a single process could account for both slow and rapid policy changes, and that relatively stable policy environments could be affected by an exogenous shock. A policy system can suffer such a shock, after which prevailing power arrangements in dominant policy groups are altered irrevocably.

A further key concept in the theory of punctuated equilibrium is that of *problem indicators*: the baseline information about an issue. These can take many forms and come from many sources, but are usually provided to policy elites in the form of second-hand data sources and presented in quantitative form (Liu, Lindquist and Vedlitz, 2011). These problem indicators may have existed for long periods before an issue comes to the attention of policy makers. A dramatic event (in Baumgartner and Jones's model, an *information shock*) is required to bring these indicators to public attention. With regard to climate change, the problem indicators include various types of climate data and information about climate impacts. The third element required to bring about rapid change in a policy environment is *information feedback*. It is not enough for problem indicators to exist or for there to be an information shock; pressure from interested actors and groups, calls for action from campaigners and politicians, and other forms of communication from claims-makers are also required to amplify the previous two stages (Jones and Baumgartner, 2005; Liu, Lindquist and Vedlitz, 2011).

Punctuated equilibrium theory seeks to explain agenda dynamics in an evolutionary way, arguing that some issues evolve slowly, while others may experience rapid change through a combination of problem indicators, information shocks, and information feedbacks. Punctuated equilibrium builds on earlier work relating to the "scope of participation" and "conflict expansion" (Schnattschneider, 1960) and competition between social groups to advance selected issues (Cobb and Elder, 1972; Dearing and Rogers, 1996). It emphasises the concept of issue competition by suggesting that policy subsystems may compete for territory and try to annexe issues that previously belonged to other policy systems. Changes in regulation, or the entry of a new strategic actor into the field, may spark a competition for dominance in a once-stable and settled policy area. This accords with findings by Nisbet on the relative success in encouraging public engagement with social issues by re-framing scientific issues as moral or emotional ones (Nisbet, 2009). In this way, punctuated equilibrium is connected to the idea of framing, in that it suggests that those with the power to define or frame an issue will wield the most influence over policy. It is also significant in that it emphasises the concept of power in issue dynamics, arguing that the power balance on policy issues may be subject to rapid change, but that power ultimately rests with those who can dominate its definition and constrain the boundaries of discourse.

The ***multiple streams*** approach (Kingdon, 1995) is similar in structure to punctuated equilibrium, in that it is based on ideas of pre-existing problems that come to policy attention because of a sudden change in circumstance. Kingdon argues that there are essentially two methods of policy prioritisation: agenda-setting, whereby issues worthy of attention are selected from the range of all possible problems, and alternative specification, whereby the set of conceivable alternatives for addressing each problem is selected (Durant and Diehl, 1989, p. 180). The former is characterised by sudden discontinuities and political actors; the latter by incremental change and nonelected members of specialised policy communities. Kingdon suggests that policy formation is best conceptualised as "multiple streams": a *problem stream* representing the real-world indicators that a problem exists; a *policy stream*, in which as range of solutions exists, awaiting a suitable problem, and a *political stream* comprised of factors that influence the body politic. These three streams flow along independently of one another until a *policy window* opens and the streams cross each other (Béland and Howlett, 2016, p. 222).

Once an issue has risen to the top of the policy agenda, it is subject to *feedback* (akin to *information feedback* in Baumgartner and Jones's approach). This feedback may come internally from government officials monitoring the operation of programmes or policies, from scientists researching in policy-relevant areas, from public opinion polls or from interest groups (Liu, Lindquist and Vedlitz, 2011, p. 407).

Multiple streams theory has been criticised for being too US-centric and insufficiently tested for relevance to other territories (Béland and Howlett, 2016). Furthermore, the vivid language and metaphors of Kingdon's original text may lead to a situation where the metaphors of streams and windows may become disconnected from the original theory (Ibid., p. 224). It has also been criticised for arguing that rapid change can take place only in the agenda-setting component of policy formation rather than in both the agenda-setting and alternative specification components. This position means it "cannot fully capture the interaction of these analytically distinct streams" (Durant and Diehl, 1989, p. 182).

Multiple streams theory draws on organisational theory, while punctuated equilibrium theory draws on evolutionary theory. Both suggest that long-standing social problems can be brought to the top of the policy agenda by means for focussing events (or information shocks) and feedbacks from interested groups, politicians, or measurements of the public mood. Media are almost entirely absent from these models. Yet it may be argued that media are the means by which these information shocks are delivered and feedback amplified. If agenda-setting theory (discussed in the subsequent section) assigns almost complete power to set the public agenda to the media, then these theories of issue dynamics downplays the media's influence. This omission of media influence, and the characterisation of the media as passive conduits in agenda systems, seems to underestimate media influence assigned by agenda-setting and theories of media effects.

In their attempt to develop of model of how social problems rise and fall, Hilgartner and Bosk (1988) proposed a *public arenas* model, with six components:

i   A dynamic process of competition among the members of a very large "population" of social problem claims;
ii The institutional arenas that serve as "environments" where social problems compete for attention and grow;
iii The "carrying capacities" of these arenas, which limit the number of problems that can gain widespread attention at one time;

iv The "principles of selection" or institutional, political, and cultural factors that influence the probability of survival of competing problem formulations;

v Patterns of interaction among the different arenas, such as feedback and synergy, through which activities in each arena spread throughout the others; and

vi The networks of operatives who promote and attempt to control particular problems and whose channels of communication crisscross the different arenas.

(Ibid., 1988, p. 56)

The public arenas in which these social problems compete for attention comprise the executive and legislative branches of government, the courts, made-for-TV movies, the cinema, the news media (television news, magazines, newspapers, and radio), political campaign organisations, social action groups, direct mail solicitations, books dealing with social issues, the research community, religious organisations, professional societies, and private foundations. Each of these arenas has a limited "carrying capacity," a concept that echoes McCombs's statement that "polling shows that the US public agenda ranges from two to six issues" at any one time (McCombs, 2004, p. 38).

The authors suggest that "principles of selection" also influence which social problems rise to the top of the agenda. These include a need for drama, novelty, the organisational characteristics of the institutions involved, and the broad cultural and political concerns of these institutions (Hilgartner and Bosk, 1988, p. 61 et seq.). To a media scholar, some of these principles (for example, novelty and drama) appear closely aligned with news values (Galtung and Ruge, 1965; Harcup and O'Neill, 2001). The public arenas model acknowledges that too much media attention can lead to saturation; high levels of coverage can "de-dramatise" a particular issue or an entire class of issues. Again, this concept is closely related to the phenomenon of "issue fatigue" identified by communications research (Djerf-Pierre, 2012a), while at the same time running counter to much of the agenda-setting literature which suggests that high levels of coverage equate to high levels of public concern (McCombs and Shaw, 1972). A final component of the public arenas model involves "feedback," communications strategies, or claims-making, which can dampen or amplify the attention given to problems in public arenas.

A key concept in the public arenas model is the idea of competition. Issues compete for limited attention in various public arenas. Advocates attempt to increase their chances of success by appealing to more than one arena or by broadening the scope of their chosen issue. Indeed, "a relatively small number of very successful social issues tend to occupy much of the space in most of the arenas" (Hilgartner and Bosk, 1988, p. 77). Furthermore, the authors argue that "the level of attention devoted to a social problem is not a function of its objective makeup alone but is determined by a process of collective definition" (Ibid., 1988, p. 68). The power of definition over an issue is again related to the framing ability of the media.

Sociologist Sheldon Ungar attempted to extend the public arenas model to suggest that social problems, especially environmental problems, have a better chance of success when they "piggyback" on real-world events (1992). These events may then trigger a *social scare*; these scares are defined as "acute episodes of collective fear that accelerate demands in the political (or related) arena" (Ibid., p. 485). When it comes to issues such as nuclear energy and climate change, Ungar argues, there is a latent dread present in the public mind; only when this dread becomes more pressing does a social scare ensue and only then do the claims-making activities of those involved with the issue gain public acceptance.

Ungar takes this concept of issue cultures a step further by suggesting that the existence of an *issue culture* is a prerequisite for the creation of a social scare (Ungar, 2014). For example, he argues, there is a long-standing issue culture around infectious diseases (Ebola, mad cow disease, SARS, swine flu), the atmosphere (the ozone hole, the greenhouse effect), and national security (the 9/11 attacks, the "War on Terror"). These "extended cultural preoccupations" (Ibid., p. 238) provide a ready-made context for similar issues as they arise. Thus, a social problem with an established issue culture is more likely to rise to the top of the public agenda because the public are already familiar with similar issues in the same culture: "The media and the public become far more receptive to claims-making that meshes with the prevailing issue culture than to claims that do not fit or resonate with it" (Ibid., p. 238).

Matthew Nisbet and Mike Huge use Downs's attention cycle model as a starting point for their attempt to construct a generalisable theory of *mediated issue development* (Nisbet and Huge, 2006). Taking the issue of plant biotechnology, the authors sketch the rise and fall of media coverage using the Downsian approach, before combining other approaches from the sociology of social problems to come up with their own model which conceptualises "several important underlying social mechanisms that drive cycles of media attention and definition to policy issues" (Ibid., 2006, p. 7). In particular, they identify and investigate four factors which, they argue, influence the progress of an issue through the attention cycle:

i   The type of policy venue where the debate takes place or is centred;
ii   The media lobbying activities of competing strategic actors as they attempt to interpret or "frame" the issue advantageously;
iii The tendency for different types of journalists to depend heavily on shared news values and norms to narrate the policy world; and
iv   The context relative to other competing issues.

(Nisbet and Huge, 2006, p. 7)

The first factor in whether an issue achieves "celebrity status" is policy venue. The authors argue that, when issues are confined to technical and scientific policy venues from which the public is excluded, media coverage is low, change occurs incrementally, and discourse is characterised by consensus. In these venues, the "scope of participation" (Schnattschneider, 1960) is also low. However, when an issue moves into administrative and more overtly political arenas, consensus is replaced by conflict, change can be non-incremental, and media coverage is increased. In regulatory policy venues, science and industry may be granted a "political monopoly" and the authority of science is defended through the use of impersonal language and technical discourse. Second, the authors consider the framing strategies of those actors trying to broaden the scope of participation of an issue and to move it from technical venues to more political ones. This approach acknowledges the importance of the power to define an issue and therefore to frame it in advantageous ways. As other researchers have noted, "framing an issue is therefore a strategic means to attract more supporters, to mobilise collective actions, to expand actors' realm of influences, and to increase their chances of winning" (Pan and Kosicki, 2001, p. 40). Schnattschneider goes even further, calling the ability to define a given issue "the supreme instrument of power" (Schnattschneider, 1960).

Nisbet and Huge also give weight to the work practices and routines of journalists in their model of mediated issue development. Just as moving from technical to political policy venues can help promote an issue, so too can changes in the kind of journalist covering it. The authors argue that, when a scientific or technical issue is covered solely by science correspondents, it tends to be framed in scientific or technical terms. However, once it begins to be covered by political correspondents, it is more likely to be presented in the strategy frame and media coverage increases (Nisbet and Huge, 2006, p. 13). The authors do not elaborate on the fourth factor in their model, "context relative to other issues," perhaps because it is self-evident that more urgent and dramatic news usually rises to the top of the news agenda. Indeed, in the opening to their paper, the authors outline a controversy over the contamination of food products with genetically modified corn. Just when public concern was at its height, the recount in Florida at the conclusion of the 2000 US presidential election came to dominate the news agenda.

The model of mediated issue development is attractive because it examines the underlying factors which produce Anthony Downs's issue attention cycle. It draws together concepts from Baumgartner and Jones's punctuated equilibrium theory (specifically, the analysis of sub-governments, their inherent biases, and relative imperviousness to outside influence) and from earlier social theorists such as Schnattschneider and Cobb and Elder (1972) on the changes in power structures that come about when new participants join the debate relating to an issue. The proposed mediated issue development theory is also attractive because it gives due acknowledgment to the role of the media in issue dynamics and because it is attuned to the norms and practices of journalists. The authors give regard to the influence of the media in general in providing the forum in which various claims-makers may be heard, but also to the individual journalist and the influence of the framing choices he or she may make as a correspondent or editor.

The media's role in social problems: theories of media effects

The theory of **agenda-setting**, namely that the media influences public perceptions of issue importance, is now almost 50 years old. The original Chapel Hill experiment (in which voters were asked about the election issues they thought were most important; their answers were compared to the media coverage of these issues, and a correlation was found) took place in 1968 (McCombs and Shaw, 1972), but the origins of the notion that audiences perceive a mediated environment rather than an actual one is even older, going back to Walter Lippmann's suggestion that the public responds to a "pseudo-environment" rather than reality or a real environment (Lippmann, 1922).

The publication of the Chapel Hill findings marked a new departure in communications research. In the 1920s and 1930s, the "hypodermic" or "magic bullet" approach had dominated, which put forward the notion that the mass media directly influenced their audiences (Scheufele and Tewksbury, 2007). This paradigm was in turn disrupted by the work of sociologist Paul Lazarsfeld of Columbia University's Bureau for Social Research in the 1940s. Lazarsfeld placed an intermediary between the press and the public, arguing that the media influenced "opinion leaders" who in turn influenced the public, a "two-step" communication process (Lazarsfeld, Berelson and Gaudet, 1948). Furthermore, Lazarsfeld contended that, in contrast to the belief of journalists, advertising executives and politicians that mass media campaigns can produce widespread changes in public opinion, the opposite is the case (cited in Katz, 1987). This strand of research emphasised the limited effects of the media and transferred power to the individual and away from the media on the basis of each individual's selective perception of the world. McCombs and Shaw disagreed with the diminution of the effect of the media, as McCombs later stated:

> ...the idea of powerful media effects expressed in the concept of agenda-setting was a better explanation for the salience of issues on the public agenda than was the concept of selective perception, which is a keystone on the idea of minimal mass media consequences.
>
> (McCombs, 2004, p. 6)

Many instances have been found of the correlation between the media's agenda and the public's agenda (McCombs, 2004, pp. 8–16), yet the agenda-setting effect is not straightforward. For example, the *time frame* is important when measuring agenda-setting effects: the effect of agenda-setting is most marked in the month after coverage appears and declines thereafter. TV news has the most immediate agenda-setting effect, but also has the shortest period before "decay" sets in. "It is primarily the accumulation of these lessons over the period of one to eight weeks that is reflected in the responses of citizen students when we enquire about the most important issues facing the nation" (McCombs, 2004, pp. 46–7). The *medium* also seems to matter, in that print has a greater agenda-setting impact than television (Ibid., p. 49). The *circumstances* in which agenda-setting occurs is important too. Agenda-setting is at its most powerful when the public requires orientation regarding an issue which is relevant to them but about which they are uncertain. The author cites the example of the Clinton-Lewinsky scandal; although media coverage was extensive and prolonged, it did not appear as a matter of importance in public opinion as it was an issue on which the public was not in need of orientation (Ibid., p. 56–7). Whether an issue is *obtrusive* or *unobtrusive* has an effect on the media's agenda-setting impact (Ibid., p. 61–2). The agenda-setting power of the media was found to be strong for unobtrusive issues of which the public has little personal knowledge, but weak for more obtrusive issues more directly experienced by individuals. For instance, Canadians followed the media agenda for unobtrusive and abstract notions such as Canadian unity but not for inflation, of which they had daily experience (Winter, Eyal and Rogers, 1982).

The media may also influence which aspect of a news event or a person was most deserving of attention. McCombs calls these "attributes," and the influence of the media in emphasising them became known as *second-level agenda-setting*. Hester and Gibson (2003, p. 74) offer a valuable definition:

> Examining the tone of news coverage, rather than simply the amount of coverage, is part of what is considered 'second-level' or 'attribute' agenda setting. This second level analyses the attributes afforded issues and individuals in news coverage, whereas traditional agenda-setting research has focused primarily on amount and placement of news coverage.

Attributes may be straightforward (e.g. a person's age, place of birth, whether they are left- or right-handed) or more complex (e.g. their political alignment; McCombs, 2004, p. 71). In selecting which attributes of an issue, or a person, to foreground, the media "not only tell us what to think, but that they tell us *how to think about* (emphasis in original) some objects" (McCombs, 2004, p. 71).

It is clear that second-level agenda-setting is similar to framing, in that it foregrounds certain aspects of an individual or issue. Indeed, the terms have been used interchangeably (Castilla, Rodríguez and País, 2014), while McCombs states that "Attribute agenda-setting explicitly merges agenda-setting theory with framing." He also argues that a frame is an attribute "because it describes the object" (McCombs, 2004, p. 88). This seems like an attempt to annexe the area of framing research as part of the territory of agenda-setting. In some definitions, framing and second-level agenda-setting are aligned (Tankard et al., 1991), while others resist the subsuming of framing into agenda-setting on the grounds that different cognitive processes are at work in each case (Scheufele, 2000). I would argue for a separation between second-level agenda-setting and framing on the grounds that framing is a more complex process. It involves not only foregrounding an aspect of an issue or individual (as takes place in second-level agenda-setting), but also involves placing the topic in a wider context and presenting it in ways that suggest both causes and solutions (Entman, 1993). Furthermore, agenda-setting (and priming, discussed in the next section) places emphasis on story selection on the part of the media, whereas framing instead focuses on the ways in which those issues are presented (Price and Tewksbury, 1997; Scheufele and Tewksbury, 2007).

In their 1987 paper, Iyengar and Kinder argued that the media not only influence public opinion as to what are the most important issues of the day but also suggest that certain issues are the ones the performance of governments, political parties, and individual politicians are to be judged against. *Priming* "changes in the standards that people use to make political evaluations" (Iyengar and Kinder, 1987, p. 63). Other definitions have refined the concept: "Priming occurs when news content suggests to news audiences that they ought to use specific issues as benchmarks for evaluating the performance of leaders and governments" (Scheufele and Tewksbury, 2007). Thus, priming as a media effect is a consequence of the agenda-setting influence of the media. Many studies have found empirical evidence of a priming effect (Miller and Krosnick, 1996), while some have found that, while an effect exists, it is minimal (Pan and Kosicki, 1997). Again, McQuail is sceptical of priming as an established theory of media effects: "Like agenda-setting, although it seems true to what is happening, it is difficult to prove in practice" (McQuail, 2010, p. 514).

Much of the academic literature comparing and analysing the differences between agenda-setting, priming, and framing discuss their operation at a psychological level (Scheufele, 1999; Scheufele and Tewksbury, 2007; Weaver, 2007), focusing on the influence each effect may have on the ease with which issues (or frames) can be brought to mind or retrieved from the memory of the individual. As noted in this and previous sections, there is ample evidence of an agenda-setting effect of the media. The effects of agenda-setting and priming have been established, yet the processes by which they take place are not entirely clear.

The importance of framing in public perceptions of social problems

Framing analysis is a common analytical framework for analysing socially constructed concepts and has become much more prevalent in recent times than agenda-setting or priming. For instance, a search of the *Communication Abstracts* database for the time frame 2001–2005 found 165 framing articles, as opposed to 43 agenda-setting studies and 25 priming studies (Weaver, 2007). The relative popularity of framing as an analytical approach may be because it "provides a strong hypothesis that an audience will be guided by journalistic frames in what it learns" (McQuail, 2010, p. 511). Framing studies are particularly prevalent within media and science communication (see Anderson, 2009, for an overview) and political science (Chong and Druckman, 2007b). Framing is the most frequently used theory in leading international journals in the field this century (Bryant and Miron, 2004). Framing's founding text is *Framing analysis: an essay on the organisation of experience* by Erving Goffman (1974) and the approach has its origins in sociology. Goffman used framing as a metaphor for "the organisation of information in everyday life" (Bowe et al., 2012, p. 158).

Many definitions of framing build on Goffman's original notion that information is organised or interpreted for the receiver. Frames have been described as "organising principles" (Reese, 2003, p. 11); a frame is "a central organising idea" (Gamson and Modigliani, 1989, p. 3); framing is "the setting of an issue within an appropriate context to achieve a desired interpretation or perspective" (de Blasio and Sorice, 2013, p. 62) which "organises the world for both journalists and those who rely on their reports" (Bowe et al., 2012, p. 158). Some scholars argue that framing is done subconsciously (Lakoff, 2009), but most agree it involves some level of conscious selection in a way "that encourages certain interpretations and discourages others" (Kitzinger, 2007, p. 134). Framing is "an inevitable process of sense-making" (Ibid.) that structures what might otherwise be random and unconnected information. Indeed, "there is no such thing as unframed information" (Nisbet, 2009, p. 15).

Framing selects some aspect of an issue for emphasis, thus encouraging audiences to think about it in a certain way. This process of selection also involves promoting "*a particular problem definition, causal interpretation, moral evaluation, and/or treatment recommendation* for the item described" (Entman, 1993, p. 52, emphasis in original). This aspect of framing – that it entails elements of problem definition, a diagnosis of causes, and a suggestion for solution (Entman, 2004) – is central to the attraction of framing as a methodological tool. In interacting with framed information, audiences may perceive what the problem is, who is to blame, and what needs to be done.

Another perspective on framing comes from Donald Schön and his collaborator Martin Rein, who consider frames as world views or paradigms to which actors in intractable policy debates are wedded. Policy development and programme implementation can be moved forward only when these actors are encouraged to reframe their standpoints (Schön and Rein, 1994). The authors refined their conceptualisation of frames in a later work, describing them as "strong and generic narratives that guide both analysis and action in practical situations" (Rein and Schön, 1996, p. 89). The authors do not give much consideration to frames found in media texts, but focus on those found in policy debates among politicians and policymakers. They put forward a model of "action frames" and "rhetorical frames": the former devolve from the observed action and the latter from crafted rhetoric in public speech and texts (Ibid., p. 92). They suggest a frame is a kind of scaffold, much like the frame of a house, on which much can be built but which remains below the surface. Such frames can be inflexible when circumstances change, just as it may be easier to demolish a house rather than adapt it to changed conditions. Their conceptualisation of framing has not been adapted much to communications or media studies, as Goffman's or Gamson's has, but their notion of a resonant narrative present in frames is useful when considering the contest by various interested parties to establish the dominant narrative in the media around climate change.

Framing is also open to criticism due to the lack of a standardised approach to detecting frames in a given text. Each new study tends to develop its own more or less unique frame set (Dahl, 2015, p. 44). Suggestions have been made for a set of generic and issue-specific frames (de Vreese, 2005), while others have pointed out the unsuitability of such a taxonomy (Porter and Hulme, 2013). Frames are usually operationally defined (Bowe et al., 2012, p. 158), meaning that for each study, a specific set of frames is created for the research project at hand. Framing researchers "have a tendency to 'reinvent the wheel'" (Nisbet, 2010, p. 46) when identifying frames. However, journalists use "generic frames" that do not depend on time, place, or subject matter (Iyengar, 1991; Neuman, Just and Crigler, 1992; Valkenburg, Semetko and de Vreese, 1999). For instance, Semetko and Valkenburg (2000) found that five frames recur in general news coverage: the responsibility, conflict, (economic) consequences, human interest, and morality frames.

Entman (1993) suggests that framing is under-theorised, while Matthes, who undertook a 15-year content analysis of framing studies in leading communications journals, found five areas for concern: the lack of transparency between the conceptualisation of the frame and its operationalisation; the lack of distinction between generic and thematic frames; the predominance of descriptive rather than theoretical studies; the lack of attention to the contribution of visual elements to the creation of frames; and the lack of reliability reporting in "a vast amount of studies" (Matthes, 2009, pp. 359–60). The diversity of approaches has "pushed the field towards incoherence" (Bowe et al.,

2012, p. 158). Some believe that scholars "often give an obligatory nod to the literature before proceeding to do whatever they were going to do in the first place" (Reese, 2007, p. 151). Standing back from the debates over methodological incoherence in framing research, McQuail questions whether news frames produce a media effect at all: "Despite the complexities, there is insufficient evidence, especially from political communications research, to demonstrate the occurrence of effects that are in line with news frames" (McQuail, 2010, p. 512). Others argue that framing effects can occur, but for them to do so, an audience is required to pay close attention to a text: "Attention to messages may be more necessary for a framing effect to occur than an agenda- setting effect. Mere exposure may be sufficient for agenda setting, but it is less likely to be so for framing effects" (Scheufele and Tewksbury, 2007).

Other perceived shortcomings of framing include a contention that it concentrates on the "micro level" of analysis and therefore may miss the real-world experience of a competitive message. Because framing concentrates its analytical focus at the level of the text – a newspaper article or television news report, for example – it "does not allow for individuals being exposed to competing frames simultaneously" (Nisbet et al., 2013, p. 767). The frame effects contained in one text may be reduced or eliminated by competing frames in other media (Sniderman and Theriault, 2004; Chong and Druckman, 2007a, 2007b; Borah, 2011).

Framing is "one of the most fertile areas of current research in journalism and mass communications" generally (Riffe, 2004, p. 2) and media coverage of climate change in particular (Schäfer et al., 2016, p. 9). In frame studies of media coverage of climate change, researchers may use deductive frames (i.e. frames derived from theory and from the existing academic literature), inductive frames (i.e. frames which emerge from the data being analysed), or a mixture of both. Some scholars suggest that all media content can be ascribed to just two deductive frames: episodic and thematic (Iyengar, 1991), while others suggest there are in fact five generic frames into which all news media content falls: the responsibility, conflict, (economic) consequences, human interest, and morality frames (Semetko and Valkenburg, 2000). A rare example of research on media coverage of climate change which takes a purely deductive approach (based on Semetko and Valkenburg's approach) is a 2010 study of Dutch and French media coverage of COPs between 2001 and 2007, which found that the consequences and responsibility frames were most common, while the conflict and human interest frames were rare (Dirikx and Gelders, 2010).

A wider set of deductive climate change frames, drawn from the literature of science communication and studies of various science-related controversies, such as nuclear power and biotechnology, has been put forward by Matthew Nisbet, who suggests eight frames applicable to climate change: *social progress* (akin to the opportunity frame in other studies), *economic development and competitiveness* (the economic frame in other studies), *morality and ethics, scientific and technical uncertainty, Pandora's Box, public accountability and governance, middle way/alternative path*, and *conflict and strategy* (Nisbet, 2009). The same author, in collaboration, has also suggested a slightly more concise typology: *economic development, morality/ethics, scientific uncertainty, Pandora's Box, public accountability*, and *conflict/strategy* frames (Nisbet and Scheufele, 2009). A more recent study proposes a typology which includes *health, security*, and *settled science* frames (O'Neill et al., 2015).

Framing is particularly attractive to students of media coverage of climate change. It may be argued that some theoretical approaches to this topic, especially those which focus closely on the text of climate change communication, operate at too micro a level to offer insight into a problem that exhibits "tremendous complexity, transcending economic, political, scientific, and social boundaries across cultures" (Bowe et al., 2012, p. 159). Nor can macro-level theories of media alone explain the media's reaction to climate change because they neglect the agency of the journalist, the news values which inform editorial decisions, newsroom cultures, and the influence of professional norms. A "meso-level" approach, which considers both the text and the context, is required. Such an approach would take into account the micro-level lexical choices of the individual journalist but also permit analysis of the broader contexts in which these choices are made. Despite the many objections to frame analysis on the basis of lack of theoretical or methodological coherence, framing continues to be attractive to media and social science researchers. As Matthew Nisbet remarks: "Framing is an unavoidable reality of the communication process, especially as applied to public affairs and policy...most successful communicators are adept at framing, whether using frames intentionally or intuitively" (2009, p. 15).

References

Anderson, A. (2009) 'Media, politics and climate change: Towards a new research agenda', *Sociology Compass*, 2(3), pp. 166–182.

Baumgartner, F.R. and Jones, B.D. (1993) *Agendas and Instability in American Politics*. Chicago, IL: University of Chicago Press.

Béland, D. and Howlett, M. (2016) 'The role and impact of the multiple-streams approach in comparative policy analysis', *Journal of Comparative Policy Analysis: Research and Practice*, 18(3), pp. 221–227.

de Blasio, E. and Sorice, M. (2013) 'The framing of climate change in Italian politics and its impact on public opinion', *International Journal of Media and Cultural Politics*, 9(1), pp. 59–69.

Borah, P. (2011) 'Conceptual issues in framing theory: A systematic examination of a decade's literature', *Journal of Communication*, 61(2), pp. 246–263.

Bowe, B. J. et al. (2012) 'Framing of climate change in newspaper coverage of the East Anglia e-mail scandal', *Public Understanding of Science*, 23(2), pp. 157–169.

Brossard, D., Shanahan, J. and McComas, K. (2004) 'Are issue-cycles culturally constructed? A comparison of French and American coverage of global climate change', *Mass Communication & Society*, 7(3), pp. 359–377.

Bryant, J. and Miron, D. (2004) 'Theory and research in mass communication', *Journal of Communication*, 54(4), pp. 662–704.

Castilla, E.B., Rodríguez, L.T. and País, E. (2014) 'Political polarization and climate change: The editorial strategies of The New York Times and El País newspapers', *Interactions: Studies in Communication & Culture*, 5(1), pp. 71–91.

Chong, D. and Druckman, J.N. (2007a) 'A theory of framing and opinion formation in competitive elite environments', *Journal of Communication*, 57(1), pp. 99–118.

Chong, D. and Druckman, J.N. (2007b) 'Framing theory', *Annual Review of Political Science*, 10(1), pp. 103–126.

Cobb, R.W. and Elder, C.D. (1972) *Participation in American Politics: The Dynamics of Agenda-Building*. Boston, MA: Allyn and Bacon.

Dahl, T. (2015) Contested science in the media: Linguistic traces of news writers' framing activity, *Written Communication*, 32(1), pp. 39–65.

Dearing, J.W. and Rogers, E.M. (1996) *Agenda Setting*. Thousand Oaks, CA: Sage Publications.

De Vreese, C.H. (2005) News framing: Theory and typology, *Information Design Journal & Document Design*, 13(1), pp. 51–62.

Dirikx, A. and Gelders, D. (2010) 'To frame is to explain: A deductive frame-analysis of Dutch and French climate change coverage during the annual UN conferences of the parties', *Public Understanding of Science*, 19(6), pp. 732–742.

Djerf-Pierre, M. (2012a) 'The crowding-out effect: Issue dynamics and attention to environmental issues in television news reporting over 30 years', *Journalism Studies*, 13(4), pp. 499–516.

Djerf-Pierre, M. (2012b) 'When attention drives attention: Issue dynamics in environmental news reporting over five decades', *European Journal of Communication*, 27, pp. 291–304.

Downs, A. (1972) 'Up and down with ecology – the issue attention cycle', in Protess, D.L. and McCombs, M.E. (eds.) *Public Interest*. London; New York: Routledge, 28, pp. 38–50.

Durant, R.F. and Diehl, P.F. (1989) 'Agendas, alternatives, and public policy: Lessons from the US foreign policy arena', *Journal of Public Policy*, 9(2), pp. 179–205.

Entman, R.M. (1993) 'Framing: Towards clarification of a fractured paradigm', *Journal of Communication*, 43(4), pp. 51–58.

Entman, R.M. (2004) *Projections of Power: Framing News, Public Opinion, and US Foreign Policy*. Chicago, IL: University of Chicago Press.

Galtung, J. and Ruge, M.H. (1965) 'The structure of foreign news. The presentation of the Congo, Cuba and Cyprus crises in four Norwegian newspapers', *Journal of Peace Research*, 2(1), pp. 64–91.

Gamson, W.A. and Modigliani, A. (1989) 'Media discourse and public opinion on nuclear power: A constructionist approach', *American Journal of Sociology*, 95(1), pp. 1–37.

Goffman, E. (1974) *Frame Analysis: An Essay on the Organization of Experience*. New York: Harper & Rowe.

Harcup, T. and O'Neill, D. (2001) 'What is news? Galtung and Ruge revisited', *Journalism Studies*, 2(2), pp. 261–280.

Hester, J.B. and Gibson, R. (2003) 'The economy and second-level agenda-setting: A time-series analysis of economic news and public opinion about the economy', *Journalism and Mass Communication Quarterly*, 80(1), pp. 73–90.

Hilgartner, S. and Bosk, C.L. (1988) 'The rise and fall of social problems: A public arenas model', *The American Journal of Sociology*, 94(1), pp. 53–78.

Howlett, M. (1997) 'Issue-attention and punctuated equilibria models reconsidered: An empirical examination of the dynamics of agenda-setting in Canada', *Canadian Journal of Political Science*, 30(1), pp. 3–29.

Iyengar, S. (1991) *Is Anyone Responsible? How Television Frames Political Issues*. Chicago, IL; London: University of Chicago Press.

Iyengar, S. and Kinder, D.R. (1987) *News That Matters: Television and American Opinion*. Chicago, IL: University of Chicago Press.

Jones, B.D. and Baumgartner, F.R. (2005) *The Politics of Attention: How Government Prioritizes Problems*. Chicago, IL: University of Chicago Press.

Jones, B.D. and Baumgartner, F.R. (2012) 'From there to here: Punctuated equilibrium to the general punctuation thesis to a theory of government information processing', *Policy Studies Journal*, 40(1), pp. 1–19.

Katz, E. (1987) 'Communications research since Lazarsfeld', *Public Opinion Quarterly*, 51(2), pp. S25–S45.

Kingdon, J.W. (1995) *Agendas, Alternatives and Public Policies*. 2nd edn. New York: Harper Collins.

Kitzinger, J. (2007) 'Framing and frame analysis', in Devereux, E. (ed.) *Media Studies: Key Issues and Debates*. London: Sage Publications, pp. 132–162.

Lakoff, G. (2009) *Why Environmental Understanding, or 'Framing,' Matters: An Evaluation of the EcoAmerica Summary Report*, Huffingtonpost.com.

Lazarsfeld, P.F., Berelson, B. and Gaudet, H. (1948) *The People's Choice: How the Voter Makes up His Mind in a Presidential Campaign*. New York: Colombia University Press.

Lippmann, W. (1922) *Public Opinion*. New York: Harcourt, Brace.

Liu, X., Lindquist, E. and Vedlitz, A. (2011) 'Explaining media and congressional attention to global climate change, 1969–2005: An empirical test of agenda-setting theory', *Political Research Quarterly*, 64(2), pp. 405–419.

Matthes, J. (2009) 'What's in a frame? A content analysis of media framing studies in the world's leading communication journals, 1990–2005', *Journalism & Mass Communication Quarterly*, 86(2), pp. 349–367.

Mazur, A. (1998) 'Global environmental change in the news', *International Sociology*, 13(4), pp. 457–472.

McCombs, M.E. (2004) *Setting the Agenda: The Mass Media and Public Opinion*. Cambridge: Polity Press.

McCombs, M.E. and Shaw, D.L. (1972) 'The agenda-setting function of mass media', *The Public Opinion Quarterly*, 36(2), pp. 176–187.

McQuail, D. (2010) *Mass Communication Theory*. 6th edn. Los Angeles, CA; London: Sage Publications.

Miller, J.M. and Krosnick, J.A. (1996) 'News media impact on the ingredients of presidential evaluations: A program of research on the priming hypothesis', in Mutz, D.C., Sniderman, P.M. and Brody, R.A. (eds.) *Political Persuasion and Attitude Change*. Ann Arbor, MI: University of Michigan Press, pp. 79–100.

Neuman, W.R., Just, M.R. and Crigler, A.N. (1992) *Common Knowledge: News and the Construction of Political Meaning*. Chicago, IL; London: University of Chicago Press.

Nisbet, M.C. (2009) 'Communicating climate change: Why frames matter for public engagement', *Environment: Science and Policy for Sustainable Development*. Taylor & Francis, 51(2), pp. 12–23.

Nisbet, M.C. (2010) 'Knowledge into action: Framing the debates over climate change and poverty', in D'Angelo, P. and Kuypers, J.A. (eds.) *Doing News Framing Analysis: Empirical and Theoretical Perspectives*. New York; London: Routledge, pp. 43–46.

Nisbet, M.C. and Huge, M. (2006) 'Attention cycles and frames in the plant biotechnology debate: Managing power participation through the press/policy connection', *The Harvard International Journal of Press/Politics*, 11(2), pp. 3–40.

Nisbet, M.C. and Scheufele, D. A. (2009) 'What's next for science communication? Promising directions and lingering distractions', *American Journal of Botany*, 96(10), pp. 1767–1778.

Nisbet, E.C. et al. (2013) 'Attitude change in competitive framing environments? Open-/closed-mindedness, framing effects, and climate change', *Journal of Communication*, 63, pp. 766–785.

O'Neill, S. et al. (2015) 'Dominant frames in legacy and social media coverage of the IPCC Fifth Assessment Report', *Nature Climate Change*, 5(4), pp. 380–385.

Pan, Z. and Kosicki, G.M. (1997) 'Priming and media impact on the evaluations of the president's performance', *Communication Research*, 24(1), pp. 3–33.

Pan, Z. and Kosicki, G.M. (2001) 'Framing as a strategic action in public deliberation', in Reese, S.D., Gandy, O. and Grant, A. (eds.) *Framing Public*

Life: Perspectives on Media and Our Understanding of the Social World. Mahwah, NJ: Lawrence Erlbaum, pp. 35–65.

Porter, K.E. and Hulme, M. (2013) 'The emergence of the geoengineering debate in the UK print media: A frame analysis', *The Geographical Journal*, 179(4), pp. 342–355.

Price, V. and Tewksbury, D. (1997) 'News values and public opinion: A theoretical account of media priming and framing', in Barnett, G.A. and Boster, F.J. (eds.) *Progress in Communication Sciences*. New York, NY: Ablex, pp. 173–212.

Reese, S.D. (2003) 'Prologue – framing public life: A bridging model for media research', in Reese, S.D., Gandy, O.H. and Grant, A.E. (eds.) *Framing Public Life. Perspectives on Media and Our Understanding of the Social World.* Mahwah, NJ: Lawrence Erlbaum, pp. 23–48.

Reese, S.D. (2007). The framing project: A bridging model for media research revisited. *Journal of communication*, 57(1), pp. 148–154.

Rein, M. and Schön, D. (1996) 'Frame-critical policy analysis and frame-reflective policy practice', *Knowledge and Policy*, 9(1), pp. 85–104.

Riffe, D. (2004) 'An editorial comment', *Journalism & Mass Communication Quarterly*, 81(Spring 2004), pp. 730–731.

Schäfer, M.S. et al. (2016) *Investigating Mediated Climate Change Communication: A Best-Practice Guide.* Research report. School of Education and Communication, Jönköping University. Available at http://hj.diva-portal. org/smash/get/diva2:961854/FULLTEXT01.pdf

Scheufele, D. (1999) 'Framing as a theory of media effects', *Journal of communication*, 49(1), pp. 103–122.

Scheufele, D.A. (2000) 'Agenda-setting, priming, and framing revisited: Another look at cognitive effects of political communication', *Mass Communication & Society*, 3(2 & 3), pp. 297–316.

Scheufele, D.A. and Tewksbury, D. (2007) 'Framing, agenda setting, and priming: The evolution of three media effects models', *Journal of Communication*, 57(1), pp. 9–20.

Schnattschneider, E.E. (1960) *The Semi-Sovereign People.* New York: Holt, Reinhart and Winston.

Schön, D.A. and Rein, M. (1994) *Frame Reflection: Toward the Resolution of Intractable Policy Controversies.* New York: Basic Books.

Semetko, H.A. and Valkenburg, P.M. (2000) 'Framing European politics: A content analysis of press and television news', *Journal of Communication*, 50(2), pp. 93–109.

Sniderman, P.M. and Theriault, S.M. (2004) 'The structure of political argument and the logic of issue framing', in Saris, W.E. and Sniderman, P.M. (eds.) *Studies in Public Opinion: Attitudes, Nonattitudes, Measurement Error, and Change.* Princeton, NJ and London: Princeton University Press, pp. 133–165.

Tankard, J., Hendrickson, L., Silberman, J., Bliss, K. and Ghanem, S. (1991) *Media frames: Approaches to conceptualization and measurement.* Paper presented to the Association for Education in Journalism and Mass Communication. Communication Theory and Methodology Division, Boston, MA.

Trumbo, C. (1996) 'Constructing climate change: Claims and frames in US news coverage of an environmental issue', *Public Understanding of Science*, 5(3), pp. 269–283.

Ungar, S. (1992) 'The rise and (relative) decline of global warming as a social problem', *The Sociological Quarterly*. Wiley Online Library, 33(4), pp. 483–501.

Ungar, S. (2014) 'Media context and reporting opportunities on climate change: 2012 versus 1988', *Environmental Communication: A Journal of Nature and Culture*, 8(2), pp. 233–248.

Valkenburg, P.M., Semetko, H.A. and de Vreese, C.H. (1999) 'The effects of news frames on readers' thoughts and recall', *Communication Research*, 26(5), pp. 550–569.

Weaver, D.H. (2007) 'Thoughts on agenda setting, framing, and priming', *Journal of Communication*, 57(1), pp. 142–147.

Winter, J.P., Eyal, C. and Rogers, A. (1982) 'Issue-specific agenda-setting the whole as less than the sum of the parts', *Canadian Journal of Communication*, 8(2), pp. 1–10.

3   The media and climate change

Introduction

Broadly, there are two main strands of media coverage of climate change research: studies which strive to understand why coverage rises and falls in particular contexts and studies which examine the nature of the coverage itself. These two approaches could also be said to be focused externally (at the social, ideological, and institutional forces affecting general media attention for climate change) and internally (at the work of the journalist, the framing, and lexical choices he or she makes, the sourcing practices he or she engages in, and the professional norms to which he or she is subject). This extensive body of research can also be divided into three distinct thematic phases: an initial phase of research which concerned itself with measuring media attention, with assessing whether the media reported climate science accurately, with ascertaining whether the scientific consensus on the anthropogenic element of climate change was reflected in the coverage, and with detecting the presence of sceptic, denialist, or contrarian voices; a second phase involving a more diverse examination of media texts through the explanatory lenses of framing theory, critical discourse analysis, and other theoretical approaches. The coverage of discreet events, rather than the broad topic of climate change, was examined in this phase (e.g. COPs, IPCC reports, and weather events). Research into social media coverage of climate change was also undertaken in this phase. And a third phase which has diversified into coverage of various constituent parts of the climate change issue. In this phase, subjects such as visual representations of climate change, media treatment of carbon, perceptions of risk, the role of celebrities, flooding, weather, transition to carbon-free societies, sustainability, landscape, and energy are examined. These phases are broadly aligned with the chronology of media coverage of climate change research, with the first phase occurring in the years after the subject became a media topic following NASA scientist James Hansen's testimony to a US Congressional hearing in 1988. The alignment of chronology with these three groups is not rigid, however; research on framing and attention levels continues to be published (e.g. Wagner and Payne, 2015), as do studies devoted to the analysis of sceptic discourses (e.g. Reed, 2016).

Theoretical approaches to understanding the media

To examine how the media have represented climate change, one must first understand how the media represents things in general. The perspectives offered by media studies, journalism studies, various ethnographic approaches, political economy, and other academic disciplines are valuable in analysing how the media work on three levels: media systems, newsroom cultures, and the sociology of the individual editor or reporter.

The *media system* in operation in a given territory at a given time will necessarily have an influence on the content of the media outlets within that system (Siebert, 1956). There have been various attempts to provide a reliable taxonomy of media systems, starting with Fred Siebert's influential work in 1956, *Four Theories of the Press*, in which four basic media systems (authoritarian, libertarian, social responsibility, and Soviet totalitarian) were proposed. A fifth, based on media ownership, was suggested (Merrill and Lowenstein, 1979) and subsequently dismissed as superfluous (Ostini and Fung, 2002) because the *Four Theories* models were based on media ownership as well as media function. Other taxonomies have been put forward: one with three categories (market, communitarian, and advancing; Altschull, 1995); one with five categories (authoritarian, Western, communist, revolutionary, and developmental; Hachten, 1981). Calling for an update of Siebert's model, Hallin and Mancini suggest three basic media models: Mediterranean/Polarised Pluralist; North/Central European or Democratic Corporatist; and North Atlantic or Liberal (Hallin and Mancini, 2004). Ireland was one of the 18 countries they analysed; it belongs, in their analysis, to the North Atlantic or Liberal model.

A basic question at the heart of these studies is how each particular kind of media system affects the performance of the media in relaying accurate information to the public. Does, for instance, the commercialisation of the media promote or inhibit this flow of information? The evidence is "fragmentary and not entirely consistent" (Hallin and Mancini, 2004, p. 279) on this point, but it is asserted that commercialisation "has encouraged the development of a globalised media culture that substantially diminishes national differences" (2004, p. 282). For others, however, the danger of the commercialisation of the media is quite clear. The flow of political information reduces in line with the level of commercialisation (Aalberg, van Aelst and Curran, 2010). Indeed, there is a strong case to be made that market-based media are structurally incapable of serving the public at all. Because the media is part of the market economy, they inevitably serve the interests of the political and economic elite (Herman and Chomsky, 1988).

Various scholars have defended the media against these charges, principally on the grounds that individual journalists have professional standards and adhere to professional norms which prevent them from slavishly following the agenda of the social elite. They are concerned with their own legitimation (Hallin, 1985) and are in fact insulated by commercialisation rather than compromised by it (Schudson, 2011, pp. 124–5).

The influence of commerce and the market also forms part of Pierre Bourdieu's critique of media systems. Bourdieu's influential concept of "the field" (any specialised form of human activity) having its own rules and forms of capital both supports and counters more political economy arguments such as those made by Herman and Chomsky and Habermas. The support comes from Bourdieu's contention that journalism is a "weak autonomous field" (Bourdieu, 2005), in that it is dependent for its survival on other fields (the economic field, for instance) by which it is inevitably influenced. However, within journalism itself, individual journalists compete for capital within the field. Thus, although the media generally may be susceptible to influence and control from economic elites, this influence may be subverted by the struggle for peer reputation and prestige of the reporters and journalists themselves (see Willig, 2012, for an overview).

Following on from theories of how media systems operate, a body of scholarship has grown around how the media in general and journalism/journalists in particular *ought* to be – **normative theories of the media**. Many lists of journalism attributes have been compiled and propounded, based squarely on the idea that a fair and free press is essential to the proper functioning of democracy (Altschull, 1995; Schudson, 2001; Deuze, 2005a, 2005b; Deuze, 2008; Schudson, 2008b; McQuail, 2010). Precisely how the press should contribute to this functioning was the subject of "one of the most instructive and heated intellectual debates of the American twentieth century" (Alterman, 2008) between journalist and political commentator Walter Lippmann and philosopher John Dewey. Lippmann (1922) believed the world was becoming too complex for lay audiences to comprehend, and the role of the media was to interpret these complexities on their behalf; Dewey (1922) believed the press should encourage the participation of the citizenry in democracy. Modern journalism has followed the Lippmann model (Schudson, 2008a), although the growth of online media has promoted a more Deweyan, participative model (Hermida et al., 2011) in which news is constructed through a conversation between journalist and citizen.

Many scholars agree that journalists should be autonomous and independent, fair and balanced, but the extent to which journalists should pursue a goal of objectivity is a matter of debate. For instance, Marx believed that adherence to a code of objectivity on the part of journalists is a barrier to change (Altschull, 1995), yet journalists themselves value objectivity as a professional norm (Schudson, 2001); indeed, it is the "chief occupational value of American journalism" (Ibid., p. 149). The American emphasis is notable; objectivity does not play such a central role in the professional norms of European media culture (Chalaby, 1996; Schudson, 2001).

Perhaps the professional norms of individual journalists could indeed mitigate against the institutional bias of the media organisations they worked for, as Michael Schudson has argued. However, there is a good deal of evidence from *ethnographic newsroom studies* to suggest that reporters and editors reproduce the bias of their employers (White, 1950; Gieber, 1964) and that newsroom culture enforces such bias (Breed, 1955; Tuchman, 1972; Glasgow University Media Group, 1976; Schlesinger, 1978; Gans, 1979). In much of this early research, there is a feeling that "if left alone, reporters will produce truth" (Reese and Ballinger, 2001, p. 651). This is a comforting notion, but one that fails to take into account the influence of news values (Galtung and Ruge, 1965; Harcup and O'Neill, 2001). It turns out that the media do not simply report the truth, but rather the truth is made to accommodate the news routines and values of the media. Furthermore, far from being mindful of professional norms and social responsibilities, journalists often rely on heuristics (Dunwoody and Griffin, 2011), a sixth sense for news, the journalistic gut feeling (Schultz, 2007).

Barbie Zelizer (2004) and Michael Schudson (2011) have provided valuable analyses of the development of sociological approaches to the study of journalism. The many approaches they describe often turn out to be ways in which journalism is weighed in the balance and found wanting. It is argued that the media systems in which journalists operate, the requirements of news itself, the inherent biases of the individual journalists all make it difficult, if not impossible, for journalists to fulfil the roles assigned to them by normative theories of the press. However, the professional norms to which individual journalists aspire may be a defence against such institutional and structural constraints, especially in cases where coverage is provided by specialist correspondents, who are seen by their peers more as "independent experts, free to make judgments" (Schudson, 2001, p. 163). Furthermore, the influence of news values – characteristics of events and topics which make them attractive to journalists – may prove more powerful than more institutional constraints.

Newsroom perspectives: professional norms, sources, and news values

Much early research into media coverage of climate change from the science communication perspective draws attention to the differing norms of science and journalism. These differences are viewed almost entirely from the point of view of the science community and contain the implication that scientific norms are the standard against which journalistic norms are measured. First, scientists point out that journalism is fast moving, while science is slow; journalism seeks unequivocal facts, while scientific progress is incremental and most scientific statements are equivocal (Bell, 1994, pp. 259–60). Journalists believe it is part of their job to be critical of sources, while scientists assign less weight to this function; journalists accept that their job has an entertainment function, but scientists do not accept this function as readily; journalists go to great lengths not to "talk down" to their audience, whereas scientists have a paternalistic attitude to media audiences (Peters, 1995, pp. 44–5). Perhaps most importantly, scientists expect journalists to support the scientist's goals, whereas journalists are neutral as to the goals of the scientific endeavour they are covering.

The work of political scientist W. Lance Bennett on professional norms was an important influence on two major works of scholarship on media coverage of climate change. Bennett argued that there were three first-order journalistic norms: political (that journalists must provide the citizenry with accurate information so as to make informed electoral choices), economic (that journalism in a market system must be efficient and profitable), and journalistic (that journalists must be fair, balanced, accurate, and objective in their reporting; Bennett, 1996). In their influential paper, Boykoff and Boykoff (2004) focused on the second-order journalistic norm of balance. Balance is variously defined as neutrality (Entman, 1989) or presenting both sides of an argument (Gans, 1979). However, simply quoting a source with an opposing view can be a way of avoiding checking the validity of the initial claim; it is a "surrogate for validity checks" (Dunwoody and Peters, 1992, p. 10). Boykoff and Boykoff's content analysis of the US prestige press from 1988 to 2002 showed that "balanced" reporting has produced a biased account of the scientific consensus. They conclude that "balanced reporting is actually problematic in practice when discussing the human contribution to global warming and resulting calls for action to combat it" (2004, p. 134). In a subsequent paper, the same authors develop their conceptualisation of journalistic norms, suggesting there are first-order norms (personalisation, dramatisation, and novelty) and second-order norms (authority bias and balance; Boykoff and Boykoff, 2007). The authors undertake a case history of US newspaper and television coverage of climate change from 1988 to 2004. They conclude that

> Adherence to the norms of dramatisation, personalisation, novelty, balance, and authority-order is part of a process that eventuates in informationally biased coverage of global warming. This informational bias has helped to create space for the US government to defray responsibility and delay action regarding climate change.
>
> (2007, p. 12)

The demonstration by Boykoff and Boykoff's 2004 paper that the standard journalistic practice of "hearing the other side" was not appropriate to media coverage of climate change and in fact produced an informational bias had a sobering effect on environmental journalists, in the US particularly (Hiles and Hinnant, 2014). Indeed, subsequent studies found that the practice of quoting sceptics had declined considerably in the US press by 2006 (Boykoff, 2007a).

Bennett's normative theories emerge from a broader sociological tradition pioneered by Herbert Gans and Robert Entman, media scholars and social theorists rather than dedicated journalism scholars. To some extent they echo earlier scholarship relating to *news values* – characteristics necessary for any event to be covered by the news media. Johan Galtung and Mari Ruge's investigation into how events become news has been called the "single piece of research that most cogently advanced a general understanding of news selection processes" (Zelizer, 2004, p. 54). Galtung and Ruge suggest 12 factors:

- *frequency* (discreet events, such as murders, are more likely to be covered than long-term social trends);
- *threshold* (events must be reasonably uncommon);

- *unambiguity* (the more easily an event can be understood, the more likely it is to be selected);
- *meaningfulness* (the more an event fits into the selector's cultural frame of reference, the more likely it is to be regarded as news);
- *consonance* (the selector may want, or predict, that something will happen; if it does, it has a greater chance of becoming news);
- *continuity* (if something is already news, that increases its chances of staying in the news);
- *composition* (an event may be included because it helps the overall balance or composition of a newspaper or broadcast);
- *reference to elite nations* (the actions of elite nations are seen as intrinsically more consequential that those of other nations);
- *reference to elite people* (famous people may be seen by news selectors as consequential);
- *reference to people* (news is presented as the result of individuals' actions rather than the result of social forces);
- *reference to something negative* (negative news is seen as more unexpected and happens over a short time span).

(Galtung and Ruge, 1965, pp. 65–71)

In a critique of Galtung and Ruge's work, Harcup and O'Neill examined over 1,200 news articles in the UK press and suggested that certain categories of story were not accounted for in the earlier taxonomy of news values. They suggest that picture opportunities, entertainment, sex, humour, reference to animals, reference to elite organisations, campaigns and promotions, and reference to something positive be added (Harcup and O'Neill, 2001). Quite apart from satisfying various news values, events and topics may need to provide dramatic stories with compelling narratives (McComas and Shanahan, 1999); this requirement for drama – which is often born of conflict over the certainty and risks of climate change or the costs or benefits of action (Nisbet, 2011) – is close to Boykoff's norms of personalisation and dramatisation (discussed earlier). It is clear that for a happening to become news, it must fit quite a rigid template of what journalists consider newsworthy.

Some environmental communications scholars have suggested that news values tend to predispose the media to rely on official *sources* (Anderson, 2009; Hansen, 2011) or to foreground conflict between scientists (Corbett and Durfee, 2004). Journalists often use government or other official sources to confer authority on their reports, while sources gain "authoritative status" in a "reciprocal utility" (Carlson, 2009, p. 530). These sources often assume the role of "primary definers" who can exercise power over the terms in which social problems are discussed. The term was coined by Stuart Hall and colleagues (1978) in a study of news coverage of crime in the UK and the associated moral panic. In *Policing the Crisis*, the authors describe how journalists continuously quote official sources in their reports. This leads to a structural bias within the news media, they argue. This level of access allows institutional sources (such as government official and law enforcement officers) to establish the initial definition of the issue, in turn setting the terms of future debate (Ibid., p. 58). Hall et al. acknowledge that this preference for official sources is not ideologically driven, but rather is based on journalistic norms and newsroom work practices. However, the practice does have the effect of reinforcing the political and economic status quo (Louw, 2005). In relation to climate change, Lisa Antilla has noted that in the 1980s, scientists were the primary definers of the issue, but that by the early 1990s, politicians had taken over this role, suggesting that climate change had ceased to be a matter of scientific evidence and had become more politically contested (Antilla, 2005).

News values are often not understood by protestors (Gavin, 2010), and scientific reports which do not contain appeals to news values such as personalisation or dramatisation may be ignored by the media (O'Neill et al., 2015). However, there is a lack of scholarship examining the impact of news values on media coverage of climate change from a journalistic perspective. Field theory has been used to explore journalists' deployment of news values (Willig, 2012), and indeed the two are related: a journalist's standing among peers is related to an ability to research and write an impactful news story, and the standard of this work is judged according to the norms of news values. This links to the Bourdieusian notion of journalistic *doxa*, "the unspoken, unquestioned, taken-for granted, understanding of the news game and the basic beliefs guiding journalistic practice" (Ibid., p. 374). However, the large preponderance of studies examining media coverage of climate change, and which allude to the influence of news values on such coverage, do so from a perspective outside the media. In other words, they examine media texts and discern the influence that news values may have had in their production. There is little acknowledgement of the role news values may have played in the commissioning of such coverage in the first place, and little understanding of how difficult a story climate change is for journalists to cover, given the newsroom cultures in which they operate and the news values by which their work is judged.

The economic pressure on the media industry may also influence the extent and quality of climate change reporting. Members of the

American Society of Environmental Journalists found that the amount of time they were given to research science subjects was insufficient and that they were expected to do general news reporting as part of their job, thus further depleting the time they were able to devote to researching their science stories. They also remarked on the decline of the number of dedicated science correspondents in the American press in general (Wilson, 2000). Alison Anderson points out that "Economic, organisation and institutional pressures means that journalists in the United Kingdom have become increasing reliant upon prepackaged information, principally from public relations (PR) professionals, industries and news agencies" (Anderson, 2011, p. 538). Such dependence was found to be "extensive" (Lewis, Williams and Franklin, 2008, p. 7), with almost half of all stories including some copy from pre-packaged information. The "carrying capacity" (Hilgartner and Bosk, 1988) of newspapers has increased over the last two decades, staffing levels have not kept pace, and cutbacks have affected many journalists who are working on the environmental beat in the UK and the US (Anderson, 2011, p. 538).

Furthermore, underfunded newsrooms may have to assign climate change stories to general news reporters. These "are usually written by general-interest journalists (given that the economic crisis has reduced the presence of specialised journalists in editorial departments), who, on top of it all, have very demanding deadlines, so the notes supplied by press offices and wire services become the main (or only) source of news" (Hoyos, 2014, p. 94). In the new digital media ecosystem, science journalists are expected to fulfil more diverse roles, with the roles of convenor, curator, civic educator, and public intellectual being added to the "legacy" role of reporter (Fahy and Nisbet, 2011). However, reduced newsroom staffing and budget cuts may make fulfilling these and other roles assigned to journalists difficult if not impossible (Beam et al., 2009). In tabloid newspapers, the pressures can be even more intense (Boykoff and Mansfield, 2008), with tabloid reporters given scant training in science and environmental reporting and expected to be a "jack of all trades" (Deuze, 2005b, p. 887). Putting general news beat reporters to work on specialist areas such as climate change can exacerbate distortions of scientific information (Anderson, 2013).

How the media have performed in covering climate change

The coverage of climate change in the prestige US print media increased dramatically in 1988, from 73 climate change stories in 1987 to 285 stories in 1988, and continued to rise in subsequent years (Wilkins, 1993, pp. 75–6). However, several researchers have identified a change in the nature of the coverage in the period 1988–1989. It was at this time that climate change became transformed from what was perceived as a scientific issue to one that was perceived as an ideological or political one. In a study of US media coverage of climate change from 1987 to 1990, Wilkins shows that, in 1987 and 1988, scientists were the sources most often quoted in the US press, whereas in 1989 and 1990, government officials were the most often quoted, with industry sources second, and scientists third (Wilkins, 1993). In her study of climate change coverage in the British press from 1985 to 2001, Anabela Carvalho discovered that scientists were the "uncontested central actors and exclusive definers" of climate change until the end of 1988 (Carvalho, 2007, p. 228). These findings align with Craig Trumbo's analysis of US coverage from 1985 to 1995, which showed that scientists were quoted less often and politicians more often as the story developed over time (Trumbo, 1996). If climate change had "arrived" as an issue in the summer of 1988, by the end of that year, it had ceased to be simply a scientific matter: "At the end of 1988, the scope of potentially necessary political, social and economic transformations necessary to address climate change started to become visible" (Carvalho, 2007, p. 229).

The study of levels of media attention to climate change is based on the premise that the public equates levels of coverage with salience and on the assumption that high levels of coverage equate to high levels of public concern on the issue. Many studies – including, in part, this one – measure levels of coverage over time and attempt to discover the underlying reasons for various rises and falls in attention levels (see Schmidt et al., 2013, for an overview). Many scholars are in agreement that media coverage of climate change gradually increased from its initial appearance on the media agenda in 1988. Some early peaks are evident in 2000 and 2001, and the coverage then builds to a high point in 2009, before dropping back to roughly 2005 levels from early 2010 (Grundmann and Scott, 2014). The International Collective on Environment, Culture and Politics project at the University of Boulder, Colorado, has been monitoring global print media coverage of climate change since 2004. Their data confirms the picture of a gradual increase up to late 2009, early 2010 followed by a steep decline. Levels of coverage have increased in the period late 2015/early 2016, but have declined again since then, and indeed even this most recent peak did not approach the levels seen in late 2009/early 2010 (Figure 3.1).

This trend has been confirmed in many territories: the United States (Boykoff, 2011), Finland (Lyytimäki and Tapio, 2009), the UK (Boykoff and Mansfield, 2008), Canada (Ahchong and Dodds, 2012),

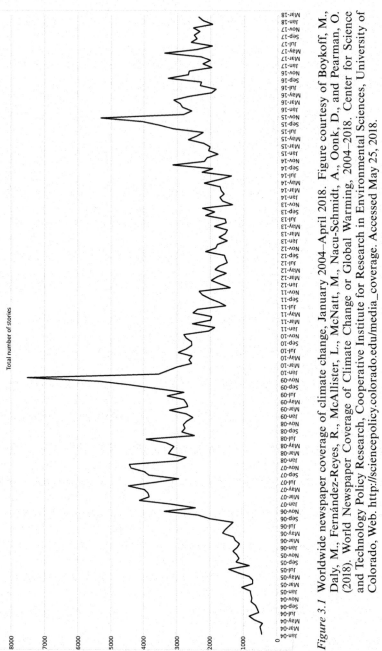

Figure 3.1 Worldwide newspaper coverage of climate change, January 2004–April 2018. Figure courtesy of Boykoff, M., Daly, M., Fernández-Reyes, R., McAllister, L., McNatt, M., Nacu-Schmidt, A., Oonk, D., and Pearman, O. (2018). World Newspaper Coverage of Climate Change or Global Warming, 2004–2018. Center for Science and Technology Policy Research, Cooperative Institute for Research in Environmental Sciences, University of Colorado, Web. http://sciencepolicy.colorado.edu/media_coverage. Accessed May 25, 2018.

Germany and France (Grundmann and Krishnamurthy, 2010), Japan (Sampei and Aoyagi-Usui, 2009), and Sweden (Shehata and Hopmann, 2012). In their key paper, Andreas Schmidt and colleagues (2013) analysed levels of climate change coverage in 27 countries from 1996 to 2010 and found that coverage had increased in all territories studied and was especially high in "carbon dependent countries with commitments under the Kyoto Protocol" (2013, p. 1233).

The various peaks and troughs in coverage are linked to political events (e.g. the 2001 withdrawal of the US from Kyoto), extreme weather events (Hurricane Katrina, August 2005), media events (the documentary *An Inconvenient Truth*, May 2006), national and international reviews and reports (the *Stern Review*, October 2006, AR4 of the IPCC, early 2007), and international conferences (the Copenhagen COP, December 2009; Grundmann and Scott, 2014, p. 223). Schmidt et al. also found that coverage peaked around COPs and the publication of IPCC reports (Schmidt, Ivanova and Schäfer, 2013, p. 1241). This may be because COPs provide a forum for political actors, to whom the news media are attracted (Galtung and Ruge, 1965; Harcup and O'Neill, 2001; Anderson, 2009), and because NGOs concentrate their activities around such conferences (Benford, 2010). Media coverage of climate change has been shown to peak at the time of COPs in the US (Boykoff and Boykoff, 2007; Boykoff and Mansfield, 2008), in Switzerland (Besio and Pronzini, 2010), Mexico (Gordon, Deines and Havice, 2010), India (Billett, 2010), Canada (Ahchong and Dodds, 2012), Japan (Sampei and Aoyagi-Usui, 2009), and Germany (Schäfer et al., 2014). The publication of Assessment Reports of the IPCC also coincides with peaks in media attention (Hulme, 2009).

There have been suggestions that the post-2009 decline follows the failure of the Copenhagen COP to arrive at an agreement, leading to a general disillusionment on the part of politicians and journalists (Lyytimäki, 2011; Schmidt, Ivanova and Schäfer, 2013), or that the decline post-2009 is due to the constriction of resources in newsrooms (Boykoff and Yulsman, 2013), which in turn has led to a reduction in the number of specialised environmental correspondents employed by media organisations (Hansen, 2011). Others have suggested that the global financial crisis "crowded out" (Djerf-Pierre, 2012) environmental news and political attention shifted to dealing with the issue (Gupta, 2010).

The presence of sceptics in coverage of climate change

From the 1980s onwards, there was a well-funded, well-organised campaign to undermine the scientific consensus on climate change.

Conservative "think tanks" played a prominent role (McCright and Dunlap, 2003; Jacques, Dunlap and Freeman, 2008; Klein, 2014). This campaign was waged primarily in the media and primarily in the US (Antilla, 2005; Oreskes and Conway, 2010). In several countries, such as France (Brossard, Shanahan and McComas, 2004), Germany (Weingart, Engels and Pansegrau, 2000), and Holland (Dirikx and Gelders, 2010), the media "exhibit less uncertainty about climate science" than in the UK and the US (Painter, 2015, p. 11). The media have been found to give undue prominence to sceptics in the US and the UK (Boykoff and Boykoff, 2004; Boykoff, 2007b; Boykoff and Mansfield, 2008), mainly due to journalists seeking to "balance" their stories by quoting a sceptic source. Some research into sceptic views present in the media concluded that contrarian or denialist views had been relegated to the margins of media discourse by the mid-2000s (Ereaut and Segnit, 2006; Boykoff, 2007b).

James Painter and Neil Gavin have pointed out some problems with this conclusion that the era of "false balance" was over: the studies claiming the sceptics had been marginalised were published in the period up to the mid-2000s "and predate very significant developments" such as the Climategate controversy, disputes over some findings in the IPCC report of 2007, and the 2009 Copenhagen COP (Painter and Gavin, 2015). Furthermore, subsequent research showed that UK tabloids continued to give space to sceptic arguments (Boykoff and Mansfield, 2008). A comparison of sceptic representations in the print media of six countries (Brazil, China, France, India, the US, and the UK) found that, while sceptic views in the first four had declined between 2007 and 2009–2010, they had in fact increased in both the US and the UK (from 18% to 34% and from 15% to 19%, respectively; Painter and Ashe, 2012). Painter and Gavin conclude that "sceptical voices were present is about one in five of all climate change articles in the main UK newspapers" between 2007 and 2011 (Painter and Gavin, 2015, p. 15). Further research focusing on two periods, 2007 and 2009–2010, "would support the view that the USA is particularly notable for the presence of sceptics who question the need for strong climate change policy proposals. It would also be true of the UK, where the GWPF [the sceptic Global Warming Policy Forum] has had a major impact in the media since its formation in November 2009" (Painter and Ashe, 2012, p. 6). In a study of broadcast and online media coverage of the Copenhagen COP of 2009, it was found that "in both the conventional and new media, contrarian claims, assumptions and lines of reasoning were still very much in evidence around the high-profile Copenhagen negotiations, or presented in such a way as to lend them undeserved authority" (Gavin and Marshall, 2011, p. 1041). Even at the prestigious British Broadcasting Corporation, where guidelines had suggested sceptics not be given equal time to proponents of the scientific consensus (Jones, 2011), the science of climate change continued to be framed as uncertain in reports from Copenhagen.

Some of this academic disagreement over whether sceptics "disappeared" from the media by the mid-2000s can be cleared up by examining the methodologies of early studies, and especially by examining their definitions of what constitutes scepticism. Much of this early, establishing work was done by American academic Maxwell Boykoff (Boykoff and Boykoff, 2004, p. 128; Boykoff, 2007a, pp. 474–5; Boykoff and Mansfield, 2008, p. 4). His methodology, broadly speaking, consisted of measuring the content of news articles against four possible portrayals of climate change: it's all due to anthropogenic causes; there is a significant anthropogenic contribution; there is a "balance" between these two; and there is little anthropogenic contribution (Painter and Gavin, 2015, p. 4). Only the second portrayal is deemed "accurate," while all others are coded as "inaccurate." This approach is narrow and reductive and does not take into account the complexity and diversity of sceptic attempts to influence media discourses on climate change. For instance, sceptics do not restrict themselves to questioning the attribution of global warming (Gavin and Marshall, 2011). In fact, there are three categories of sceptic (Rahmstorf, 2004): *trend sceptics*, who argue that global warming is not happening at all; *attribution sceptics*, who concede global warming is happening but deny the anthropogenic contribution; and *impact sceptics*, who accept the human contribution but suggest climate impacts may be beneficial or that climate models are not robust enough and/or question the necessity for policy interventions (Painter and Ashe, 2012, p. 3). (A shorthand version is: "it's not happening; it's happening but it's not us; it's happening and it's us, but there's nothing we can do.")

Thus, scepticism goes far beyond questioning the science of climate change, although sceptics do question the integrity of the satellite data, the usefulness of computer modelling, and claim that warming is due to natural variability or solar activity (Oreskes and Conway, 2010). Trend sceptics may also direct attention to a medieval warming period that cannot be attributed to CO_2 emissions or to an alleged "pause" in warming since 1998. Other sceptics ignore the scientific evidence altogether, suggesting instead that climate science is itself a conspiracy; that individual scientists dare not confess this for fear of destroying their careers; that there is a rigorously enforced orthodoxy akin to a religion; that the IPCC is politicised; and that climate change is actually a means by which socialists can implement anti-capitalist programmes (see Booker and North, 2007; Lawson, 2008; Booker, 2009 for a range of sceptic positions and Monbiot 2011 for a refutation).

There is some evidence to suggest that the presence of sceptics in newspaper coverage of climate change is linked to ideological stance of the news organisation in question (Carvalho, 2005, 2007; Carvalho and Burgess, 2005). These three studies analysed the period 1985–2000 and concluded that, of the three newspapers examined (the right-leaning *Times* and the left-leaning *Independent* and *Guardian*), the sceptic stance of the *Times* became strongly apparent by the mid-1990s. Others note that the *Daily Mail* coverage of climate change reflected the scientific consensus least and suggest that "a key element shaping the difference may be the politically conservative stance of the newspaper" (Boykoff and Mansfield, 2008, p. 4). A study of commentary in News Corporation outlets in the UK, the US, and Australia revealed a tendency to emphasise uncertainty and to frame consensus views as motivated by political correctness or adherence to orthodoxy, while sceptics were portrayed by News Corp outlets as courageous dissenters (McKnight, 2010). It has also been found that, while other major news organisations "overwhelmingly reflected the consensus view on the reality and causes of climate change" between 2009 and 2010, the *Wall Street Journal* did not (Nisbet, 2011, p. 57).

It is clear that early studies of scepticism in media coverage of climate change focused on a narrow definition of scepticism; thus, when the sceptics ceased to focus simply on denials of climate science, they ceased to be recorded as sceptics at all in these early coding methodologies. Once the complexities and nuances of scepticism were more fully understood, and their tactics uncovered, a more thorough content analysis of media could be undertaken. This showed that, far from being banished to the margins of media coverage of climate change, they had in many cases increased their presence. In the US and the UK at least, they are very far from being a "dead norm" (Boykoff, 2007a).

Air, water, and pollution: the disaggregation of climate change

Recent scholarship of media coverage of climate change has moved away from the treatment of the issue as a homogenous entity. Perhaps partly in acknowledgement of the complexity of climate change, and partly because of the contribution of earlier research, scholars have focused on its constituent parts. Recent studies have considered, for instance, the social media coverage of individual COPs (O'Neill et al., 2015), of flooding (Gavin, Leonard-Milsom and Montgomery, 2011; Escobar and Demeritt, 2014; Devitt and Neill, 2017), online comments sections on climate change (Koteyko, Jaspal and Nerlich, 2013), climate change coverage in science magazines (Nielsen and Schmidt Kjærgaard, 2011), scientist bloggers (Thorsen, 2013), letters to the editor on climate change

(Young, 2013), images used to illustrate climate change (O'Neill, 2013; Rebich-Hespanha et al., 2015), and representations of coal (Bacon and Nash, 2012), carbon (Feldpausch-Parker et al., 2015; McNally, 2015), and fracking (Jaspal and Nerlich, 2014). The academy has moved on from establishing whether media coverage of climate change accepts anthropogenic global warming as happening and has begun to consider what a post-climate change world might look like.

However, certain characteristics remain persistent. The perspective of journalists is all too rare, a recent survey of elite environmental correspondents notwithstanding (Hiles and Hinnant, 2014). In her review of media coverage of climate change trends, Alison Anderson suggests that "research involving in-depth interviews with editors and reporters would provide important insights into the factors impeding or enhancing climate change coverage" (Anderson, 2009, p. 176). The view from the other side of the media fence, from those wishing to increase, influence, or dominate media coverage of climate change, is also lacking. Anderson adds: "Studies involving in-depth interviews with news sources would provide us with a greater understanding of competition to control the issue and the behind-the-scenes factors influencing patterns of reporting" (2009, p. 176). Anderson echoes a similar suggestion made by Max and Jules Boykoff that "... interviews with journalists would better situate and extract explanations as to why journalists continue to adhere to the norm of balanced reporting on the issue of global warming at certain times, and not in others. Also, future studies could integrate macro-structural analysis with the micro-process analysis featured here. Furthermore, future work could delineate partial predictive influences on the production of 'balanced' coverage, of global warming, or divergence from it, in order to more finely texture explanations of this media coverage" (Boykoff and Boykoff, 2004, p. 134). It is possible to argue that the academic field has failed to shake off its origins in science communication and remains overly concerned with scientific accuracy and portrayals of science; the difference is that now such concern relates to representations of issues such as carbon capture and storage and wind turbines rather than climate change itself.

References

Aalberg, T., van Aelst, P. and Curran, J. (2010) 'Media systems and the political information environment: A cross-national comparison', *The International Journal of Press/Politics*, 15(3), pp. 255–271.

Ahchong, K. and Dodds, R. (2012) 'Anthropogenic climate change coverage in two Canadian newspapers, the Toronto Star and the Globe and Mail, from 1988 to 2007', *Environmental Science and Policy*, 15(1), pp. 48–59.

Alterman, E. (2008) 'Out of print: The death and life of the American newspaper', *The New Yorker*, March 31, 2008 issue. Available at https://www.newyorker.com/magazine/2008/03/31/out-of-print

Altschull, J.H. (1995) *Agents of Power: The Media and Public Policy*. White Plains, New York: Prentice Hall.

Anderson, A. (2009) 'Media, politics and climate change : Towards a new research agenda', *Sociology Compass*, 2(3), pp. 166–182.

Anderson, A. (2011) 'Sources, media, and modes of climate change communication: The role of celebrities', *Wiley Interdisciplinary Reviews: Climate Change*, 2(4), pp. 535–546.

Anderson, A. (2013) *Media, Culture and the Environment*. Abingdon, Oxon, UK: Routledge.

Antilla, L. (2005) 'Climate of scepticism: US newspaper coverage of the science of climate change', *Global Environmental Change*, 15(4), pp. 338–352.

Bacon, W. and Nash, C. (2012) 'Playing the media game: The relative (in)visibility of coal industry interests in media reporting of coal as a climate change issue in Australia', *Journalism Studies*, 13(2), pp. 37–41.

Beam, R.A., Weaver, D.H. and Brownlee, B.J. (2009) 'Changes in professionalism of U.S. journalists in the turbulent twenty-first century', *Journalism & Mass Communication Quarterly*, 86(2), pp. 277–298.

Bell, A. (1994) 'Media (mis)communication on the science of climate change', *Public Understanding of Science*, 3, pp. 259–275.

Benford, R.D. (2010) 'Framing global governance from below: Discursive opportunities and challenges in the transnational social movement Era', in Bjola, C. and Kornprobst, M. (eds.) *Arguing Global Governance: Agency, Lifeworld and Shared Reasoning*. London; New York: Routledge, pp. 67–84.

Bennett, W.L. (1996) 'An introduction to journalism norms and representations of politics', *Political Communication*, 13(4), pp. 373–384.

Besio, C. and Pronzini, A. (2010) 'Unruhe und Stabilität als Form der massenmedialen Kommunikation über Klimawandel', in *Der Klimawandel*. VS Verlag für Sozialwissenschaften, pp. 283–299.

Billett, S. (2010) 'Dividing climate change: Global warming in the Indian mass media', *Climatic Change*, 99(1), pp. 1–16.

Booker, C. (2009) *The Real Global Warming Disaster: Is the Obsession with 'Climate Change' Turning Out to be the Most Costly Scientific Blunder in History?* London: Continuum.

Booker, C. and North, R. (2007) *Scared to Death: From BSE to Global Warming—How Scares Are Costing Us the Earth*. London: Continuum.

Bourdieu, P. (2005) 'The political field, the social science field, and journalistic field', in Benson, R. and Neveu, E. (eds.) *Bourdieu and the Journalistic Field*. Cambridge: Polity Press, pp. 29–47.

Boykoff, M.T. (2007a) 'Flogging a dead norm? Newspaper coverage of anthropogenic climate change in the United States and United Kingdom from 2003 to 2006', *Area*, 39(4), pp. 470–481.

Boykoff, M.T. (2007b) 'From convergence to contention: United States mass media representations of anthropogenic climate change science', *Transactions of the Institute of British Geographers*, 32(4), pp. 477–489.

Boykoff, M.T. (2011) *Who Speaks for the Climate? Making Sense of Media Reporting on Climate Change.* Cambridge: Cambridge University Press.

Boykoff, M.T. and Boykoff, J.M. (2004) 'Balance as bias: Global warming and the US prestige press', *Global Environmental Change*, 14(2), pp. 125–136.

Boykoff, M.T. and Boykoff, J.M. (2007) 'Climate change and journalistic norms: A case-study of US mass-media coverage', *Geoforum*, 38(6), pp. 1190–1204.

Boykoff, M.T. and Mansfield, M. (2008) '"Ye Olde Hot Aire": Reporting on human contributions to climate change in the UK tabloid press', *Environmental Research Letters*, 3, p. 024002.

Boykoff, M.T. and Yulsman, T. (2013) 'Political economy, media, and climate change: Sinews of modern life', *Wiley Interdisciplinary Reviews: Climate Change*, 4(5), pp. 359–371.

Breed, W. (1955) 'Social control in the newsroom: A functional analysis', *Social Forces*, 33(17), pp. 326–335.

Brossard, D., Shanahan, J. and McComas, K. (2004) 'Are issue-cycles culturally constructed? A comparison of French and American coverage of global climate change', *Mass Communication & Society*, 7(3), pp. 359–377.

Carlson, M. (2009) 'Dueling, dancing, or dominating? Journalists and their sources', *Sociology Compass*, 3(4), pp. 526–542.

Carvalho, A. (2005) 'Representing the politics of the greenhouse effect: Discursive strategies in the British media', *Critical Discourse Studies*, 2(1), pp. 1–29.

Carvalho, A. (2007) 'Ideological cultures and media discourses on scientific knowledge: Re-reading news on climate change', *Public Understanding of Science*, 16(2), pp. 223–243.

Carvalho, A. and Burgess, J. (2005) 'Cultural circuits of climate change in UK broadsheet newspapers, 1985–2003', *Risk analysis*, Wiley Online Library, 25(6), pp. 1457–1469.

Chalaby, J. (1996) 'Journalism as an Anglo-American invention. A comparison of the development of French and Anglo-American journalism', *European Journal of Communication*, 3, pp. 303–326.

Corbett, J.B. and Durfee, J.L. (2004) 'Testing public (un) certainty of science media representations of global warming', *Science Communication*, 26(2), pp. 129–151.

Deuze, M. (2005a) 'Popular journalism and professional ideology: Tabloid reporters and editors speak out', *Media, Culture & Society*, 27(6), pp. 861–882.

Deuze, M. (2005b) 'What is journalism? Professional identity and ideology of journalists reconsidered', *Journalism*, 6(4), pp. 442–464.

Deuze, M. (2008) 'The professional identity of journalists in the context of convergence culture', *Observatorio (OBS*)*, 2(4), pp. 103–117.

Devitt, C. and Neill, E.O. (2017) 'The framing of two major flood episodes in the Irish print news media: Implications for societal adaptation to living with flood risk', *Public Understanding of Science*, 26(7), pp. 872–888.

Dewey, J. (1922) 'Public opinion', *The New Republic*, 30 (May 3) pp. 286–288.

Dirikx, A. and Gelders, D. (2010) 'To frame is to explain: A deductive frame-analysis of Dutch and French climate change coverage during the annual UN conferences of the parties', *Public Understanding of Science*, 19(6), pp. 732–742.

Djerf-Pierre, M. (2012) 'The crowding-out effect: Issue dynamics and attention to environmental issues in television news reporting over 30 years', *Journalism Studies*, 13(4), pp. 499–516.

Dunwoody, S. and Griffin, R.J. (2011) 'Judgmental heuristics and news reporting', in Gowda, R. and Fox, J.C. (eds.) *Judgments, Decisions and Public Policy*. Cambridge: Cambridge University Press, pp. 177–198.

Dunwoody, S. and Peters, H.P. (1992) 'Mass media coverage of technological and environmental risks: A survey of research in the United States and Germany', *Public Understanding of Science*, 1(2), pp. 199–230.

Entman, R. (1989) *Democracy Without Citizens: Media and the Decay of American Politics*. New York; London: Oxford University Press.

Ereaut, G. and Segnit, N. (2006) 'Warm words: How are we telling the climate story and can we tell it better?', *Public Policy Research*, 2006(August), p. 32.

Escobar, M.P. and Demeritt, D. (2014) 'Flooding and the framing of risk in British broadsheets, 1985–2010', *Public Understanding of Science*, 23(4), pp. 454–471.

Fahy, D. and Nisbet, M.C. (2011) 'The science journalist online: Shifting roles and emerging practices', *Journalism*, 12(7), pp. 778–793.

Feldpausch-Parker, A.M. et al. (2015) 'Spreading the news on carbon capture and storage: A state-level comparison of US media', *Environmental Communication*, 7(3), pp. 336–354.

Galtung, J. and Ruge, M.H. (1965) 'The structure of foreign news. The presentation of the Congo, Cuba and Cyprus crises in four Norwegian newspapers', *Journal of Peace Research*, 2(1), pp. 64–91.

Gans, H.J. (1979) *Deciding What's News: A Study of CBS Evening News, NBC Nightly News, Newsweek, and Time*. Evanston, IL: Medill School of Journalism Visions of the American Press, Northwestern University Press.

Gavin, N.T. (2010) 'Pressure group direct action on climate change: The role of the media and the web in Britain – A case study', *British Journal of Politics and International Relations*, 12, pp. 459–475.

Gavin, N.T., Leonard-Milsom, L. and Montgomery, J. (2011) 'Climate change, flooding and the media in Britain', *Public Understanding of Science*, 20(3), pp. 422–438.

Gavin, N.T. and Marshall, T. (2011) 'Mediated climate change in Britain: Scepticism on the web and on television around Copenhagen', *Global Environmental Change*, 21(3), pp. 1035–1044.

Gieber, H. (1964) 'News is what newspapermen make it', in Dexter, L.A. and White, D.M. (eds.) *People, Society and Mass Communications*. New York: Free Press, pp. 173–182.

Glasgow University Media Group (1976) *Bad News*. London: Robert Kennedy Publishing (RKP).

Gordon, J.C., Deines, T. and Havice, J. (2010) 'Global warming coverage in the media: Trends in a Mexico city newspaper', *Science Communication*, 32, pp. 143–170.

Grundmann, R. and Krishnamurthy, R. (2010) 'The discourse of climate change: A corpus-based approach', *Critical Approaches to Discourse Analysis across Disciplines*, 4(2), pp. 113–133.

Grundmann, R. and Scott, M. (2014) 'Disputed climate science in the media: Do countries matter?', *Public Understanding of Science*, 23(2), pp. 220–235.

Gupta, J. (2010) 'A history of international climate change policy climate change policy', *Wiley Interdisciplinary Reviews: Climate Change*, 1(5), pp. 636–653.

Hachten, W. (1981) *The World News Prism: Changing Media. Clashing Ideologies*. Ames: Iowa State University Press, pp. 60–77.

Hall, S. et al. (1978) *Policing the Crisis: Mugging, the State and Law and Order*. London: Macmillan.

Hallin, D.C. (1985) 'The American news media: A critical theory perspective', in Forester, J. (ed.) *Critical Theory and Public Life*. Cambridge, MA: The MIT Press, pp. 121–146.

Hallin, D.C. and Mancini, P. (2004) *Comparing Media Systems: Three Models of Media and Politics*. Cambridge: Cambridge University Press.

Hansen, A. (2011) 'Communication, media and environment: Towards reconnecting research on the production, content and social implications of environmental communication', *International Communication Gazette*, 73(1–2), pp. 7–25.

Harcup, T. and O'Neill, D. (2001) 'What is news? Galtung and Ruge revisited', *Journalism Studies*, 2(2), pp. 261–280.

Herman, Edward S. and Chomsky, N. (1988) *Manufacturing Consent: The Political Economy of the Mass Media*. New York: Pantheon Books.

Hermida, A., Domingo, D., Heinonen, A., Paulussen, S., Quandt, T., Reich, Z., Singer, J.B. and Vujnovic, M. (2011) 'The active recipient: Participatory journalism through the lens of the Dewey-Lippmann debate', *International Symposium on Online Journalism*, 1(2), pp. 1–21.

Hiles, S.S. and Hinnant, A. (2014) 'Climate change in the newsroom: Journalists' evolving standards of objectivity when covering global warming', *Science Communication*, 36, pp. 428–453.

Hilgartner, S. and Bosk, C.L. (1988) 'The rise and fall of social problems: A public arenas model', *The American Journal of Sociology*, 94(1), pp. 53–78.

Hoyos, G. (2014) 'No more great expectations: Media response to Obama's climate action plan', *Interactions: Studies in Communication & Culture*, 5(1), pp. 93–106.

Hulme, M. (2009) *Why We Disagree About Climate Change: Understanding Controversy, Inaction and Opportunity*. Cambridge: Cambridge University Press.

Jacques, P.J., Dunlap, R.E. and Freeman, M. (2008) 'The organisation of denial: Conservative think tanks and environmental scepticism', *Environmental politics*, 17(3), pp. 349–385.

Jaspal, R. and Nerlich, B. (2014) 'Fracking in the UK press: Threat dynamics in an unfolding debate', *Public Understanding of Science*, 23(3), pp. 348–363.

Jones, S. (2011) *BBC Trust Review of Impartiality and Accuracy of the BBC's Coverage of Science*. London.

Klein, N. (2014) *This Changes Everything: Capitalism vs The Climate*. London; New York: Penguin.

Koteyko, N., Jaspal, R. and Nerlich, B. (2013) 'Climate change and "climategate" in online reader comments: A mixed methods study', *Geographical Journal*, 179(1), pp. 74–86.

Lawson, N. (2008) *An Appeal to Reason: A Cool Look at Global Warming*. London: Duckworth Overlook.

Lewis, J.M.W., Williams, A. and Franklin, R.A. (2008) 'A compromised fourth estate? UK news journalism, public relations and news sources', *Journalism Studies*, 9(1), pp. 1–20.

Lippmann, W. (1922) *Public Opinion*. New York: Harcourt, Brace.

Louw, E. (2005) *The Media and Political Process*. Thousand Oaks, CA: Sage.

Lyytimäki, J. (2011) 'Mainstreaming climate policy: The role of media coverage in Finland', *Mitigation and Adaptation Strategies for Global Change*, 16, pp. 649–661.

Lyytimäki, J. and Tapio, P. (2009) 'Climate change as reported in the press of Finland: From screaming headlines to penetrating background noise', *International Journal of Environmental Studies*, 66(6), pp. 723–735.

McComas, K. and Shanahan, J. (1999) 'Telling stories about global climate change measuring the impact of narratives on issue cycles', *Communication Research*. Sage Publications, 26(1), pp. 30–57.

McCright, A.M. and Dunlap, R.E. (2003) 'Defeating Kyoto: The conservative movement's impact on U.S. climate change policy', *Social Problems*, 50(3), pp. 348–373.

McKnight, D. (2010) 'A change in the climate? The journalism of opinion at News Corporation', *Journalism*, 11, pp. 693–706.

McNally, B. (2015) 'Media and carbon literacy: Shaping opportunities for cognitive engagement with low carbon transition in Irish media, 2000–2013', *Razon y Palabra*, September (91), pp. 2000–2013.

McQuail, D. (2010) *Mass Communication Theory*. 6th edn. Los Angeles, CA; London: Sage Publications.

Merrill, J.C. and Lowenstein, R.L. (1979) *Media, Messages, and Men: New Perspectives in Communication*. New York: Longman.

Monbiot, G. (2011) 'The superhuman cock-ups of Christopher Booker', *Guardian.co.uk*, 13 October.

Nielsen, K.H. and Schmidt Kjærgaard, R. (2011) 'News coverage of climate change in nature news and science now during 2007', *Environmental Communication*, 5(1), pp. 25–44.

Nisbet, M.C. (2011) *Climate Shift: Clear Vision for the Next Decade of Public Debate*. Report. Available at http://climateshiftproject.org/report/climate-shift-clear-vision-for-the-next-decade-of-public-debate/

O'Neill, S. (2013) 'Image matters: Climate change imagery in US, UK and Australian newspapers', *Geoforum*, 49, pp. 10–19.

O'Neill, S. et al. (2015) 'Dominant frames in legacy and social media coverage of the IPCC fifth assessment report – supplementary material', *Nature Climate Change*, 2, pp. 1–9.

Oreskes, N. and Conway, E.M. (2010) 'Defeating the merchants of doubt', *Nature*, 465(7299), pp. 686–687.

Oreskes, N. and Conway, E.M. (2010) *Merchants of Doubt*. London: Bloomsbury Publishing.

Ostini, J. and Fung, A.Y.H. (2002) 'Beyond the four theories of the press: A new model of national media systems', *Journalism*, 5(1), pp. 41–56.

Painter, J. (2015) *Media Representations of Uncertainty About Climate Change*. University of Westminster.

Painter, J. and Ashe, T. (2012) 'Cross-national comparison of the presence of climate scepticism in the print media in six countries, 2007–10', *Environmental Research Letters*, 7(4), p. 044005.

Painter, J. and Gavin, N.T. (2015) 'Climate skepticism in British newspapers, 2007 – 2011', *Environmental Communication*, 10(4), pp. 37–41.

Peters, H.P. (1995) 'The interaction of journalists and scientific experts: Co-operation and conflict between two professional cultures', *Media, Culture and Society*, 17, pp. 31–48.

Rahmstorf, S. (2004) 'The climate sceptics', Report by the *Potsdam Institute for Climate Impact Research* for Munich Re, pp. 76–82. Available at http://www.pik-potsdam.de/~stefan/Publications/Other/rahmstorf_climate_sceptics_2004.pdf

Rebich-Hespanha, S. et al. (2015) 'Image themes and frames in US print news stories about climate change', *Environmental Communication*. Taylor & Francis, 9(4), pp. 491–519.

Reed, M. (2016) '"This loopy idea": An analysis of UKIP's social media discourse in relation to rurality and climate change', *Space and Polity*. Taylor & Francis, 20(2), pp. 226–241.

Reese, S.D. and Ballinger, J. (2001) 'The roots of a sociology of news: Remembering Mr. Gates and social control in the newsroom', *Journalism & Mass Communication Quarterly*, 78(4), pp. 641–658.

Sampei, Y. and Aoyagi-Usui, M. (2009) 'Mass-media coverage, its influence on public awareness of climate-change issues, and implications for Japan's national campaign to reduce greenhouse gas emissions', *Global Environmental Change*, 19(2), pp. 203–212.

Schäfer, M.S. et al. (2014) 'What drives media attention for climate change? Explaining issue attention in Australian, German and Indian print media from 1996 to 2010', *International Communication Gazette*, 76(2), pp. 152–176.

Schlesinger, P. (1978) *Putting 'Reality' Together*. London; New York: Methuen.

Schmidt, A., Ivanova, A. and Schäfer, M.S. (2013) 'Media attention for climate change around the world: A comparative analysis of newspaper coverage in 27 countries', *Global Environmental Change*, 23(5), pp. 1233–1248.

Schudson, M. (2001) 'The objectivity norm in American journalism*', *Journalism*, 2(2), pp. 149–170.

Schudson, M. (2008a) 'The "Lippmann-Dewey Debate" and the invention of Walter Lippmann as an anti-democrat 1986–1996', *International Journal of Communication*, 2, pp. 1031–1042.

Schudson, M. (2008b) *Why Democracies Need an Unlovable Press*. Cambridge: Polity Press.

Schudson, M. (2011) *The Sociology of News*. 2nd edn. New York; London: W. W. Norton.

Schultz, I. (2007) 'The journalistic gut feeling', *Journalism Practice*, 1(2), pp. 190–207.

Shehata, A. and Hopmann, D. (2012) 'Framing climate change: A study of US and Swedish press coverage of global warming', *Journalism Studies*, 13(2), pp. 175–192.

Siebert, F.S. (1956) *Four Theories of the Press: The Authoritarian, Libertarian, Social Responsibility, and Soviet Communist Concepts of What the Press Should Be and Do.* Champaign: University of Illinois Press.

Thorsen, E. (2013) 'Blogging on the ice: Connecting audiences with climate-change sciences', *International Journal of Media & Cultural Politics*, 9(1), pp. 87–101.

Trumbo, C. (1996) 'Constructing climate change: Claims and frames in US news coverage of an environmental issue', *Public Understanding of Science*, 5(3), pp. 269–283.

Tuchman, G. (1972) 'Objectivity as strategic ritual: An examination of newsmen's notions of objectivity', *American Journal of Sociology*, 77(4), p. 660.

Wagner, P. and Payne, D. (2015) 'Trends, frames and discourse networks: Analysing the coverage of climate change in Irish newspapers', *Irish Journal of Sociology*, 2, pp. 1–24.

Weingart, P., Engels, A. and Pansegrau, P. (2000) 'Risks of communication: Discourses on climate change in science, politics, and the mass media', *Public Understanding of Science*, 9(3), pp. 261–283.

White, D.M. (1950) 'The gatekeeper: A case study in the selection of news', *Journalism Quarterly*, 27, pp. 383–389.

Wilkins, L. (1993) 'Between facts and values: Print media coverage of the greenhouse effect, 1987–1990', *Public Understanding of Science*, 2(1), pp. 71–84.

Willig, I. (2012) 'Newsroom ethnography in a field perspective', *Journalism*, 14(3), pp. 372–387.

Wilson, K.M. (2000) 'Drought, debate, and uncertainty: measuring reporters' knowledge and ignorance about climate change', *Public Understanding of Science*, 9(1), pp. 1–13.

Young, N. (2013) 'Working the fringes: The role of letters to the editor in advancing non-standard media narratives about climate change', *Public Understanding of Science*, 22(4), pp. 443–459.

Zelizer, B. (2004) *Taking Journalism Seriously: News and the Academy.* Thousand Oaks, CA: Sage.

4 Media coverage of climate change in Ireland

Introduction

The data and analysis presented in this chapter is based on a content analysis of climate change coverage in seven national newspapers over the period January 2007 to February 2016. A search of the Lexis Nexis database of Irish newspaper titles for the phrases "climate change" or "global warming" or "greenhouse effect" was carried out, producing a corpus of 12,865 articles. This corpus was reduced by (i) excluding shorter articles (fewer than 500 words; n = 7,059) and (ii) selecting every 10th story (n = 706). This reduced corpus was then analysed for the presence of frames. The frame typology was based on two influential sets of frames, one suggested by Nisbet as relevant to scientific topics in the media (Nisbet, 2009) and the other relevant to coverage of climate change more specifically (O'Neill et al., 2015). The coverage of climate change in the Irish print media was also compared to the coverage of all news during the period. The analysis shows that Irish media's coverage of climate change was 0.843% of total news coverage over the period, which compares to a global average of 0.62% between 1996 and 2010 (Schmidt, Ivanova and Schäfer, 2013).

Ireland has been relatively neglected as a subject territory for the study of media coverage of climate change. It has "only recently emerged as a distinct field of inquiry" (Fox and Rau, 2016, p. 1). The first two pieces of research into media coverage of climate change in Ireland were published in 2013; the first was the publication by Andreas Schmidt and colleagues of a cross-national study of media attention levels for climate change from 1996 to 2010. Ireland was one of 27 countries studied, and the *Irish Times* was the only paper whose coverage was recorded. Ireland conformed to the broad international trends: coverage peaked at about 4% of all news coverage in late 2009 (Schmidt, Ivanova and Schäfer, 2013). However, as this research has found, the *Irish Times*, especially in the 1996–2010 period, published greater levels of climate change stories than any other title, and therefore it may not be representative of Irish media generally. Also, in 2013, sociologist Gerard Mullally and his colleagues presented some data on newspaper coverage of climate change to a conference in University College Cork. Coverage in the *Irish Times, Irish Independent,* and *Irish Examiner,* along with 13 regional titles, was studied for the period 2008 to June 2013. The *Irish Times* coverage, which had been close to 500 stories in late 2009, dropped dramatically to fewer than 100 by 2013; the two other national titles did not approach the peak in coverage shown by the *Irish Times,* and their coverage, while it did decrease post-2009, did not suffer such a dramatic decline. Of the 13 provincial papers studied, only the *Corkman* and the *Kerryman* devoted notable levels of coverage to the issue (Mullally et al., 2013).

The levels of attention for climate change in the broadcast media in Ireland is also a neglected area of academic research. A single, exploratory study of RTE's coverage of the issue from October 2010 to October 2013 was undertaken at the request of the RTE Audience Council. This research concluded that the national broadcaster's coverage was "infrequent, sporadic and clustered around a small number of topical areas." The report also found that there was a decline in the number of climate change items broadcast from October 2010 to October 2011, a period during which the broadcaster left the post of environment correspondent unfilled. In the main TV news programmes, climate change was seen as an international rather than an Irish story and climate change was not linked to any topical stories. However, sceptical views were not present in the coverage studied (Cullinane and Watson, 2014, p. 19).

Levels of climate change coverage in three Irish newspapers (the *Irish Independent, Irish Times,* and *Sunday Business Post*) from 1997 to 2012 formed part of the data for a study by Wagner and Payne on the representation of climate change in the Irish media. This study confirmed earlier findings that Ireland's newspaper coverage rises and falls at times of international conferences and major scientific reports, and that levels of coverage declined steeply in late 2009/early 2010. The authors conclude that the Irish print media present a narrow, ideological view of climate change, one that is supportive of the country's political and economic elites (Wagner and Payne, 2015). Another study, by Brenda McNally, which focuses on the Irish media's representation of low carbon transition and decarbonisation and is therefore tangentially related to media coverage of climate change, examined coverage in a wide range of titles: the *Irish Times, Irish Independent, Irish*

Examiner, Sunday Business Post, Mirror and *Sunday Mirror, Sunday Tribune, Sunday Independent,* and *Irish Daily Mail.* Again, coverage of these climate change-related topics (low carbon transition, decarbonisation) peaks in late 2009, and again, the *Irish Times* publishes a far greater number of articles on these topics than the other titles (212 articles, compared to 52 in the *Irish Examiner,* the next highest; McNally, 2015).

An overview of the scholarship of media coverage of climate change in Ireland has recently been provided by an entry on Ireland in the Oxford Research Encyclopaedia of Climate Science (Fox and Rau, 2016). The authors survey the literature mentioned earlier, concluding that (i) coverage of climate change has steadily increased since 1997; (ii) there was a steep "post-crash drop" in late 2009; (iii) the *Irish Times* has a "track record of providing the most coverage"; and (iv) that coverage is largely driven by events (2016, p. 8). They further find that coverage presents climate change as a challenge that must be dealt with from within the existing institutional and economic system, and that broadsheet newspaper coverage relies heavily on the optimistic discourse of ecological modernisation (2016, p. 9).

A climate change frame formed part of a frame analysis of Irish media coverage (*Irish Times, Irish Independent, Irish Examiner*) of flooding, in which the authors noted that the media relied on descriptive reporting and emphasised humanitarian responses to individuals rather than long-term adaptation policies (Devitt and Neill, 2017). More recently, a climate change-related study of Irish media coverage in the *Irish Independent, Irish Times, Irish Examiner, Sunday Independent,* and *Irish Daily Mail* of the Papal encyclical on climate change, *Laudato Si': On Care for Our Common Home,* examined frames used by the Irish print media. The encyclical was published in June 2015 and received widespread media coverage. The study found that the *morality or ethics* frame dominated, while the *settled science* frame was also prominent, with the *political or ideological struggle* frame relegated to third place (Robbins, 2017). A report on Irish media coverage of climate change, including print, social, broadcast, and online media, is in preparation (Culloty et al., forthcoming).

The academic literature on media coverage of climate change in Ireland is evidently somewhat threadbare. It may be argued that, in the case of Ireland, the academy has bypassed certain stages in the process of analysing media coverage of climate change, stages that have been undertaken with regard to other countries. Whereas the general scholarship of media coverage of climate change began with concern over levels of attention and scientific accuracy, these areas have not heretofore been addressed with regard to Ireland. Further-more, the range of academic disciplines represented elsewhere, such as communications, media, journalism, and science communications studies, have not contributed to the study of Ireland's media coverage of climate change. Of the substantive research undertaken on Ireland, the influence of sociology and public policy (Wagner and Payne, 2015; Devitt and Neill, 2017) is apparent. These scholars use media data to make broader points about political economy and social struc-tures rather than the media itself and its influence on public engage-ment with climate change. In considering representations of carbon (McNally, 2015) and of flooding (Devitt and Neill, 2017), the literature relating to Ireland's media coverage of climate change has left other, more central questions unanswered. This work aims to address some of these shortcomings: first, by measuring media attention in print media in seven national titles over a period of over nine years, and second, by presenting a more media-focused perspective and analysis, emphasising the scholarship of media and journalism studies, and pre-senting the viewpoints of working journalists, politicians, and media advisors for the first time.

Levels of print media attention for climate change in Ireland

Irish print media coverage of climate change follows well-established international patterns: it peaks around international climate con-ferences and the release of international climate science reports (see Figure 4.1). Coverage gradually increases from 2007 onwards, with distinct peaks in November/December 2007, December 2009, and December 2015. The November (196 stories) and December (300 stories) 2007 peak coincides with (i) the publication of the Synthesis Report of the IPCC's AR4 on November 16, 2007, and (ii) COP 13 in Bali from December 3 to 15, 2007.

In their study of 27 countries from 1997 to 2009, Schmidt and his colleagues found that the average percentage of total news coverage concerning climate change was 0.60%. The authors found that coun-tries which has already suffered floods, heatwaves, storms, or other climate-related loss events (as recorded by the Climate Risk Index) had higher levels of media attention for climate change, but that countries likely to suffer future negative impacts, as forecast by their placing on the DARA Vulnerability Factor, did not show increased levels of cover-age. Countries whose economies depend to a greater extent on carbon-intensive industries show an increased level of climate change coverage

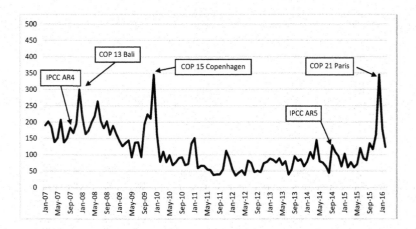

Figure 4.1 Number of climate change articles published in seven Irish national newspapers, January 2007–February 2016. *Based on a search of the LexisNexis database for seven Irish national newspapers for the dates shown. Search terms: "global warming" OR "climate change" OR "greenhouse effect".*

(Schmidt, Ivanova and Schäfer, 2013). Ireland has a low Climate Risk Index score (Sönke et al., 2015) and a low to moderate exposure to climate vulnerability (DARA Climate Vulnerability Monitor, 2012).

The authors note that, over the 27 countries they studied, two distinct periods of media attention for climate change could be identified. In the 1997–2000 period, coverage was relatively low, although it had increased somewhat from the early 1990s. In the period 2004–2009, a larger increase was recorded, what the authors call a "clear shift" (Ibid., p. 1240). Their findings for Ireland show that coverage of climate change was 0.27% of total news coverage for the 1997–2000 period (against an average over 27 countries of 0.20%); 0.51% for the period 2001–2005 (average = 0.29%) and 1.82% for the period 2006–2009 (average = 1.26%) (Ibid., p. 1241). Thus, Ireland recorded above average levels of coverage for each of the three periods identified by Andreas Schmidt and his colleagues. However, it should be noted that Schmidt and his colleagues' analysis is based on levels of coverage in the *Irish Times* only. As discussed in a subsequent section, the *Irish Times* has consistently covered climate change more extensively than other Irish newspapers, and therefore these findings cannot be said to be representative of Irish newspaper coverage more generally.

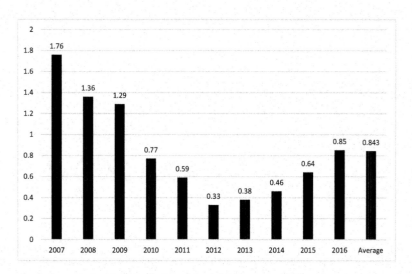

Figure 4.2 Climate change stories as a percentage of total news coverage by year. *Total news coverage established by a series of LexisNexis searches, leaving search term field blank, thus returning all news items published during dates searched.*

The findings of this research align with those of Schmidt and his colleagues with regard to the period up to 2009: Irish newspaper coverage of climate change accounts for 1.76% of total news coverage in 2007, before falling slightly to 1.36% and 1.29% in 2008 and 2009, respectively (see Figure 4.2). The post-2009 decline seen in the total levels of news coverage is also starkly evident when the coverage of climate change is examined as a percentage of total news coverage, falling to 0.77% in 2010, 0.59% in 2011, and 0.33% in 2012. My research finds that news coverage mentioning climate change comprises 0.843% of total news coverage over the entire time period studied. Although direct comparisons with Schmidt and his colleagues' findings are not possible because the time periods under examination differ, some remarks may be made about the time periods when the two pieces of research overlap. Schmidt and his colleagues note that climate change coverage reached its highest levels in the 2006–2009 period and that Ireland's climate change coverage comprised 1.82% of total news coverage during that period. My research finds that, for the somewhat shorter period of 2007–2009, the equivalent figure is 1.47%. The difference can be explained by the fact that Schmidt and his colleagues recorded coverage from the *Irish Times* only, whereas my research records coverage from seven national newspaper titles.

In considering how individual newspaper titles compare when it comes to their coverage of climate change, it is evident that the *Irish Times* publishes far more climate coverage than its rivals; however, it is also clear that the others, particularly the *Irish Independent*, are closing the gap. Taking the three most significant peaks in coverage as points of comparison, in December 2007, the *Irish Times* published 143 climate stories compared to 60 by the *Irish Independent*, 42 by the *Irish Examiner*, 21 by the *Sunday Independent*, 20 by the *Sunday Business Post*, and 10 by the *Sunday Tribune*. The *Irish Times* (143) published nearly as many climate stories as the other publications combined (157). In December 2009, the *Irish Times* published 184 climate stories compared to 60 by the *Irish Examiner*, 54 by the *Irish Independent*, 24 by the *Sunday Business Post*, 13 by the *Sunday Tribune*, and 10 by the *Sunday Independent*. In this instance, the *Irish Times*' total of 184 was more than all the other newspapers combined (161). In December 2015, the *Irish Times* total was 143, while the other newspapers (*Irish Independent* 72; *Irish Examiner* 69; *Irish Daily Mail* 28; *Sunday Independent* 33) amounted to 202. It total, over these three peak periods, the *Irish Times*'s story total of 470 compares to a combined total of 520 for all the other publications. This finding is aligned with other research: the *Irish Times* was also the "dominant news organisation" reporting on low-carbon transition and decarbonisation between 2000 and 2013 (McNally, 2015). There is other evidence that the *Irish Times* published significantly more climate change-related articles than other Irish print titles (Mullally et al., 2013; Wagner and Payne, 2015; Robbins, 2016).

It is also evident that the dominance of the *Irish Times* when it comes to climate change coverage is declining. In December 2009, it published 184 climate stories, compared to 60 from the *Irish Examiner* and 54 from the *Irish Independent*, its nearest rivals, a difference of 60 stories; in December 2015, the *Irish Times* published 143 stories, just two stories more than the total of 141 published by its two closest competitors. In April 2014, the *Irish Times* published 48 climate-related stories, but both the *Irish Independent* (40) and the *Irish Examiner* (30) were close behind in the extent of their own coverage.

The relative decline of coverage levels in the *Irish Times* and the relative increase seen in the *Irish Independent* may be influenced by staffing issues at the two newspapers. Frank McDonald, the long-serving environment editor of the *Irish Times*, retired in June 2015. For some time before his retirement, he contributed mostly feature articles and was less concerned with reporting day-to-day environmental news. Up to 2007, McDonald had been assisted by an environment correspondent. This post was left vacant when Liam Reid left to become an advisor to the Green Party Minister John Gormley. With regard to the national broadcaster RTÉ, the post of environment correspondent was left vacant from 2010 to 2014 following the departure of Paul Cunningham to another post. As it has been found that the media tend to cover issues and events already in the media (Galtung and Ruge, 1965; Harcup and O'Neill, 2001; Denham, 2014), the absence of a dedicated environment correspondent at the national broadcaster meant that an important intermedia agenda-setting effect was absent. Cunningham's replacement was given the role of covering the environment and agriculture (RTE, 2014). The organisation, in fact, placed "agriculture" ahead of "environment" in the job title of this new role, and the appointee spoke at some length about the food sector, the farming backgrounds of his parents, and the "great prospects for increased output and jobs" in the agriculture sector, before mentioning global warming (Ibid.). Others have pointed out the "inherent contradiction" of joining these two journalistic portfolios into the same role (Cunningham, personal interview, April 24, 2015) given that agriculture contributes almost 30% of the country's total emissions (Environmental Protection Agency, 2013). More recently, Kevin O'Sullivan stepped down as editor of the *Irish Times* and was appointed to the post of Environment Editor. Meanwhile, at the *Irish Independent*, Paul Melia was appointed environment correspondent in February 2008. (I have interviewed these journalists concerning climate change as a journalistic topic, and their views are presented and analysed in the next chapter.)

A comparison between Ireland and Europe is possible by comparing the data compiled for this research with that of the International Collective on Environment, Culture and Politics (ICECaPs) based at the Co-operative Institute for Research in Environmental Sciences (CIRES) at the University of Colorado at Boulder. The CIRES project monitors 52 newspaper sources across seven regions using the LexisNexis, Proquest, and Factiva databases (McAllister et al., 2017). The metric used by CIRES is average number of climate change stories per publication (i.e. the total number of climate change stories for a territory or region is recorded, and the total is then divided by the number of sources). I have used the same methodology to arrive at a similar measurement for Ireland. In Figure 4.3, it is evident that Irish print coverage of climate change follows the broad patterns of European coverage, exhibiting similar peaks and troughs, but at a lower level of overall coverage. However, it should be noted that the CIRES dataset for Europe is heavily UK-centric. The papers tracked by CIRES for Europe are the *Times* and *Sunday Times*, the *Sun*, the *Observer*, the *Guardian*, the *Daily Mail* and *Mail on Sunday*, the *Daily Mirror* and

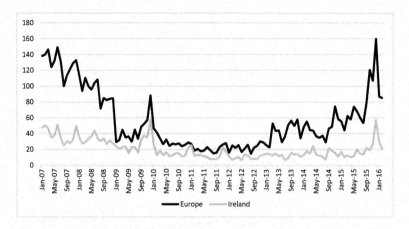

Figure 4.3 Comparison of Irish and European coverage of climate change.

Sunday Mirror, the *Daily Telegraph* and *Sunday Telegraph*, the *Financial Times*, the *Independent*, the *Scotsman* and *Scotsman on Sunday* (all UK), *El Pais* (Spain), the *Irish Times* (Ireland), and the *Sofia Echo* (Bulgaria). Several important territories, such as France, Germany, Italy, and the Scandinavian countries, are absent from the CIRES dataset.

How Irish newspaper coverage of climate change is framed

One of the criticisms of framing studies is the absence of a uniform approach that each one creates a framing typology ab initio to suit the research topic at hand (Dahl, 2015), that the frames are operationally defined (Bowe et al., 2012), and that framing researchers exhibit a tendency to "reinvent the wheel" (Nisbet, 2010, p. 46). Some of this debate on how best to operationalise frame analysis involves arguments as to whether generic frames (Semetko and Valkenburg, 2000) or "episodic" frames (Iyengar, 1991) are best suited to providing replicable results. Tankard recommends the "list of frames" approach, which identifies 11 framing mechanisms and has, he claims, the advantage of being replicable, of taking the subjectivity out of the identification of frames, of being reliable, and of allowing theory building and theory testing (Tankard, 2001, p. 101). Tankard and others (e.g. Nisbet and Scheufele, 2009) concede that, while using established frame typologies can help bring coherence to a fragmented field, there will always be a need to allow issue-specific frames emerge from the data. Tankard states that "...frames and frame indicators must be discovered and defined for each new topic under investigation. This step seems unavoidable, since framing deals fundamentally with the differences in the ways particular stories are presented" (2001, p. 102). The authors of a "best-practice" guide for researchers into media coverage of climate change recommend that scholars "look for existing conceptual tools" while also pointing out that generic approaches are best suited to issue-comparison studies, while issue-specific approaches are best suited to in-depth analysis (Schäfer et al., 2016, p. 5–10).

Accordingly, I have used a combination of two existing frame typologies here, while also allowing new frames to emerge from the data. The broad coding schema derives from typologies put forward by Matthew Nisbet (2009) and Saffron O'Neill and colleagues (2015). Nisbet, in an influential essay on the importance of frames for public engagement, suggested a typology of frames relevant to media analyses of climate change: *social progress, economic development and competitiveness, morality and ethics, scientific and technical uncertainty, Pandora's Box, public accountability and governance, middle way/alternative path*, and *conflict and strategy*. In their supplementary material supplied with their research paper, O'Neill and her colleagues provide a thorough examination of frames in existing studies before putting forward the following typology: *settled science, uncertain science, political or ideological struggle, disaster, opportunity, economic, morality and ethics, role of science, security*, and *health* (O'Neill et al., 2015). In addition to these frames commonly found in science-related policy debates, two other frames specific to the issue of climate change in an Irish context were detected in the data: a *communitarian/cosmopolitan* frame and an *agriculture* frame (Table 4.1).

A total of 1,643 frames were found in the articles. The Political or Ideological Contest frame was the most dominant, with 399 instances, or 24.28% of the total number of frames. Other frames strongly represented are settled science (238 instances; 14.48% of all frames), the economic frame (215 instances, 13.08%), the disaster frame (191 instances, 11.62%) and the policy or technical frame (184 instances, 11.19%). Frames that were not prominent in the print media coverage of climate change include the agriculture frame (51 instances, 3.1% of all frames), the morality or ethics frame (148 instances, 9%), the opportunity frame (111 instances, 6.75%), and the contested science frame (94 instances, 5.72%). The presence of the domestication or communitarian frame is almost negligible, with just 12 instances, or 0.73% of all frames recorded (Figure 4.4).

Table 4.1 Typology of frames coded

Frame	Contains
Political or ideological contest	References to summits, conferences, and talks, to climate change as a political issue, to the stances on climate change of political parties or politicians, to political posturing in advance of climate talks, to political point-scoring and jockeying for position on the issue of climate change, to assessments as to who is "winning" or "losing" in the political battle to implement climate policies, to the "game" of climate negotiation, to climate change as a battle between elites, and explicit references to a "left versus right" conflict regarding the implementation of climate policies.
Policy or technical	References to measurements, records, or assessments which are policy-neutral and which do not suggest, recommend, or imply any particular course of action.
Morality or ethics	References to the moral imperative of dealing with climate change, to the impact of inaction on future generations, to climate change generally as a religious issue, to climate change as related to humanity's stewardship of the planet, to the impact of climate change on those who have done least to cause it (climate justice), or to explicit references to fairness, justice, or equity.
Opportunity	References to climate change as an economic or business opportunity, to positive impacts or consequences of climate change mitigation and adaptation, to economic benefits of energy efficiency, to strategic opportunities for Ireland in developing green technologies, to the benefits of reducing dependency on imported fossil fuels, and to opportunities in replacing fossil fuel with renewables.
Agriculture	References to the impacts of climate change on agriculture, to the contribution of agriculture to Ireland's total emissions, to reactions to suggestions that these emissions be reduced, to the cultivation of forestry as a carbon sink, to the reduction of meat in human diet, to land use, to CAP reform or talks, and to food security.
Settled science	References to the reality of climate change and to the necessity of doing something about it, to the science of climate change, to the publication of scientific reports (including IPCC reports), to the issuing of climate data, to specific ways in which individuals or governments can undertake mitigation or adaptation measures, and to the measured impacts of climate change (e.g. flooding, rising sea levels, species depletion, crop failure, famine, temperature rise).

(Continued)

Frame	Contains
Contested science	References to climate change as not happening or being due to natural causes, suggestions that any initiatives to mitigate climate change or to reduce individual or sectoral emissions are misguided, references which confuse weather and climate, which suggest that the science of climate change is contested, that climate scientists are in error or may change their minds, which cite the University of East Anglia email theft as evidence of a conspiracy among climate scientists, which contain dismissive or sarcastic dismissals of environmental campaigners or politicians.
Disaster	References to climate change as an unavoidable disaster or as a looming apocalypse, to the catastrophic impacts of climate change (which do not contain references to the possibility of mitigating climate change, adapting to it, or to a combination of both), to the impact of climate change on specific species, or parts of the world, or to exclusively negative economic impacts.
Domestication or communitarianism	References to Ireland's emissions as a percentage of global emissions, to Ireland's emissions targets, to Ireland's negotiating strategies in climate change talks, to the impacts of climate change on Ireland specifically, and to the minimal impact on global emissions any mitigation measures Ireland may take are coded as featuring this frame.
Economic	References to the economic cost of mitigation or adaptation, to market-based solutions to climate change, to entrepreneurial activity in the climate sector, to the effects of climate change on various areas of economic activities (car manufacturing, aviation, etc.), and to the cost of energy production.

When we come to look at the framing of climate change in each newspaper title, it is apparent that the framing is relatively evenly spread between the titles. The *Irish Times* leads in each category, simply because it publishes more climate change coverage than its competitors. It is more useful to consider the percentage of each paper's coverage coded to a particular frame. When the data is presented in this manner, some interesting preferences are apparent: the *Sunday Business Post* and the *Sunday Tribune* have the highest percentage instance of the economic frame (19.67% and 19.01%, respectively), while the *Irish Daily Mail* records the highest percentage for the contested science frame (16.67%, compared to 9.65% in the *Irish Independent*, the next highest total recorded). The *Irish Examiner*

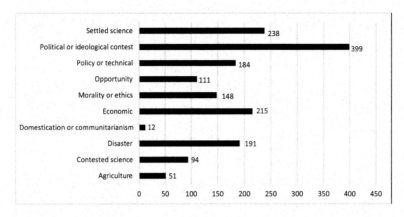

Figure 4.4 Frequency of frames in Irish print media coverage of climate change.

Table 4.2 Percentage of newspaper title coverage devoted to each frame

Title/Frame	A	O	SS	CS	E	P/T	D	M/E	PIC
Irish Times	1.93	7.2	16.43	4.73	11.49	11.92	12.24	9.56	24.49
Irish Independent	4.82	7.46	12.28	9.65	15.35	14.04	6.58	6.14	23.68
Irish Examiner	5.93	7.41	12.59	0.74	14.07	7.41	19.26	13.33	19.26
Irish Daily Mail	2.78	8.33	5.56	16.67	11.11	8.33	5.56	22.22	19.44
Sunday Tribune	3.28	1.64	11.48	4.92	19.67	14.75	16.39	4.92	22.95
Sunday Business Post	4.93	7.04	14.79	6.34	19.01	8.45	7.04	6.34	26.06
Sunday Independent	6.25	4.17	8.33	4.17	14.88	10.42	20.83	2.08	29.17

A = agriculture; O = Opportunity; SS = Settled Science; CS = Contested Science; E = Economic; P/T = Policy or Technical; D = Disaster; M/E = Morality or Ethics; PIC = Political or Ideological Contest.

Note: the Domestication or Communitarian frame was also coded for, but only 15 instances, less than 1% of all frames, were recorded, and therefore this frame is not included in this schema.

(19.26% of frames) and the *Sunday Independent* (20.83%) present much of their coverage through the disaster frame. The *Sunday Independent* and the *Irish Daily Mail* record considerably lower percentages for the settled science frame (8.33% and 5.56%, respectively, compared to 16.43% in the *Irish Times*). The *Irish Daily Mail* records a high score of 22.22% for the morality or ethics frame, far more than the second highest score for this frame, the *Irish Examiner* (13.33%). The political or ideological contest frame is also relatively evenly represented in all seven newspaper titles, with the *Sunday Independent* (29.17%) favouring this frame slightly more than the *Sunday Business Post* (26.06%), the *Irish Times* (24.49%), the *Irish Independent* (23.68%), and the *Sunday Tribune* (22.95%). The *Irish Examiner* (19.26%) and the *Irish Daily Mail* (19.44%) recorded the fewest instances of the political or ideological contest frame in their coverage (Table 4.2).

Political framings dominate Ireland's climate change coverage

The political or ideological contest frame dominates Irish print media coverage of climate change. It is found in 265 (37.5%) climate change stories. There are 399 instances of this frame within the corpus of 706 articles, accounting for 24.28% of all frames. It is the dominant frame in 199 articles (28.1%) and appears as a secondary frame in 200 other stories (28.3%).

Given that the timeframe of this study encompasses an eventful period in Irish politics, the prevalence of this frame is not surprising. The issue of climate change was an important part of discussions between the Green Party and Fianna Fail following the May 2007 general election, formed a central part of the programme for government agreed upon between the Greens, Fianna Fail and the Progressive Democrats in June 2007 (Ahern, Harney and Sargent, 2007), and was central to the policies promoted by the two Green Party ministers in the 27th and 28th governments of Ireland (Minihan, 2015).

The political frame is quite evenly represented in all newspaper titles. The lowest number of references coded to this frame occurs in the *Irish Examiner*, where it accounts for 19.26% of all frames recorded, while the *Sunday Independent* records the highest frequency, at 29.17% of all frames recorded. Frames were recorded in two ways: the dominant frame (which appeared in the article's first paragraph and/or was the main framing device used by the author) was recorded, and other, secondary frames (if any) present in the articles were also coded. Thus, it is possible to see which secondary frames accompanied the political frame most often. For instance, the *Irish Times* published far more stories dominated by the political frame (121) than any other newspaper, while the *Irish Examiner* (9), the *Irish Daily Mail* (3), and the *Sunday Tribune* (5) published very few. In stories dominated by the political frame, the secondary frames coded included the economic (34% of instances), morality (19%), policy (13%), contested science (12%), disaster (10%), opportunity (4%), settled science (4%), and agriculture (3%) frames.

The relatively strong presence of a secondary framing of contested science (12%) in political coverage is also noteworthy. Examples of this combination occur in reports of US electoral politics and reports of Dáil debates on the climate issue. For example, a report on the views of prospective Republican presidential candidate Fred Thompson in the *Irish Times* on July 28, 2007, is primarily a political story, but contains a reference to the actor's views on climate change, which he equated with a belief that the Earth is flat. A parliamentary report in the *Sunday Business Post* on July 6, 2008, noted that TDs (members of the Irish parliament) failed to distinguish between weather and climate, and that one TD remarked that his mother blamed the launch of the Sputnik spacecraft in 1957 for changes in the climate.

In looking more closely at the 399 instances of political framing in the corpus of 706 climate change articles, it is possible to identify several subtopics: the particular political arena or subject matter to which the coded text refers. The subtopics identified are outlined later, while Table 4.2 shows the number of instances for each found in politically framed articles:

i   **Climate change as political proxy**: reference to a belief in the anthropogenic element of climate change, or to the existence of climate change at all, as an indicator of more general political orientation;

ii **EU regulations**: references to the political claims-making in relation to the setting of EU regulatory target, such as CO_2 emissions targets and exhaust emissions levels for motor vehicles;

iii **EU-level politics**: references to the politics of EU climate and other policy, to Ireland's stance in relation to EU policy, to EU policy in relation to diplomatic and economic orientation towards the US and the UN, and to the Nice and Lisbon Treaty referendum campaigns;

iv **G7, G8 and G20 summits**: references to political statements and positioning in advance of and during international policy and economic summits of developed nations;

v   **Ireland Greens**: references to policies proposed by the Green Party of Ireland, to statements made by its members, and to negotiations in relation to the formation of a coalition government between the Fianna Fáil, Progressive Democrat, and Green parties following the 2007 general election in Ireland;

vi **Ireland party politics**: references to domestic political debate and claims-making in relation to climate policy, to assessment of party performance, and critiques of ministers and party leaders.

vii **Ireland regulation**: text concerning announcements of and reaction to regulatory policies proposed in Ireland, presented in a political frame without references to party politics;

viii **Politics of other territories**: references to climate politics in foreign news coverage. The single instance of this subtopic is found in an article concerning electoral politics in Australia published in the *Sunday Tribune* on December 2, 2007;

ix **UK politics**: references to climate policy and politics in the UK. The two instances of this subtopic concern UK security in the face of threats posed by climate change (*Irish Times*, March 14, 2007) and a meeting between the UK prime minister and prospective US presidential candidates at which climate change was discussed (*Irish Times*, April 18, 2008);

x **UN climate summits**: references to political manoeuvring in advance of and during UN Conferences of the Parties (COPs);

xi **UN-level politics**: references to statements made to the UN General Assembly referring to climate change or statements relating to climate change made by the UN Secretary-General.

xii **US domestic politics**: references to debates internal to the US on climate change, to the issue as a polarising one for evangelicals in the US, and to proposed domestic mitigation measures; and

xiii **US international politics**: references to the US policy position in advance of and during international climate summits, to US-EU relations as they relate to climate change, and to US influence on global affairs (Table 4.3).

Table 4.3 Subtopics in politically framed coverage

Topic	References
Climate change as political proxy	2
EU regulations	4
EU-level politics	21
G7, G8, G20 summits	5
Ireland Greens	18
Ireland party politics	35
Ireland regulation	4
Politics other territories	1
UK politics	2
UN climate summits	8
UN-level politics	3
US domestic politics	7
US international politics	14

Subtopics relating to domestic politics (57 instances) and international politics (60) are evenly represented. The international politics total comprises 25 references to EU-related topics and 35 references to broader global politics. The strong representation of outward-facing coverage aligns with the earlier finding of this research that domestication and communitarian perspectives on climate change are largely absent from Ireland's coverage of the issue.

An example of the framing of climate change as a contest for advantage among domestic political actors occurs in a report of public forum on the topic organised by the Labour Party and Friends of the Earth in the EU Parliament offices in Dublin on June 24, 2009. The forum was addressed by then Labour Party leader Eamon Gilmore, who argued that the Taoiseach should take personal responsibility for climate change. Mr Gilmore appears to acknowledge the gravity of the climate change situation, on the one hand, suggesting that a national response is required, but he appears to be seeking to score advantage over a political rival by making him personally answerable on Ireland's lack of climate action rather than advancing any climate policy himself:

> Mr Gilmore said yesterday the gravity of the situation demanded a major national response and it was the responsibility of Taoiseach Brian Cowen to provide the leadership behind that response. There is historical precedence. When there are major national issues that need to be addressed like the Northern Ireland peace process, the European Union, social partnership during the economic crisis in the 1980s, the political responsibility is transferred to the Taoiseach. Climate change targets need now to be seen in the same way as the peace process in Northern Ireland. The Taoiseach must take responsibility for it, he said.
>
> (*Irish Times*, June 25, 2009)

Issues relating to climate change such as energy independence, CO_2 emissions targets, and proposals to tax carbon coded to this frame are presented in the context of their impact on political popularity rather than their likely climate impact. In this example, nuclear power is presented in exclusively party-political terms, while wider environmental and societal contexts are not presented:

> Eamon Ryan wants a public debate on nuclear power. The Communications and Energy Minister was responding to the British government's backing for a new generation of nuclear power stations to secure the UK's energy supplies. "The debate on nuclear

is part of a wider debate on where we are going to get our energy from in the future," Ryan said. He suggested that the issue be taken up by the Oireachtas committee on climate change and energy security. But before asking people to consider the merits of a debate on nuclear power, Ryan should check the views of the larger party in the coalition government. Bertie Ahern has already rejected the nuclear power option. He aligned himself emphatically to the anti-nuclear camp last February at a Fianna Fáil conference. At the same meeting in Galway, the then energy minister Noel Dempsey said there was "no proposal before government or contemplated by government to change" the statutory ban on generating electricity by nuclear means which had been in place for almost 30 years.

(*Sunday Tribune*, January 13, 2008)

Thus, climate change is portrayed as a policy area in which political advantage may be gained or in which an opponent can be put at a disadvantage. Concern for the climate itself, or the benefits or otherwise of particular policies, appear secondary to concerns of political point-scoring.

There are several instances of journalists seeking to delegitimise attendance at climate change conferences and other events abroad on the basis of cost and CO_2 emissions incurred in travel. One such report concerns a trip by the then Minister for the Environment Phil Hogan to the Earth Summit (the UN Conference on Sustainable Development in Rio de Janeiro). The report describes the trip as "controversial" and foregrounds the costs involved:

Mr Hogan and his department have spent almost €92,000 travelling to Europe and the rest of the world since he took office in March last year. The most expensive jaunt so far was a €28,847 trip to Durban, South Africa, to attend the COP 17 Climate Change Conference in Durban in December last year. Mr Hogan's delegation stayed six nights in the beach front Blue Water Hotel. Flights for the department's controversial three-day trip to the UN Conference on Sustainable Development in Rio earlier this summer cost €7,347 for four of the members of the mission who attended. Accommodation costs for the group were the subject of much discussion between department officials and Ireland's Brazilian embassy and the final hotel bill for the trip has not yet been finalised. However, it was revealed that six members of the group dined out on the taxpayers' expense clocking up a €842.66 in one of the city's famous steak houses.

(Sunday Independent, July 19, 2012)

Other similar reports concerned trips by councillors to a range of events, including climate change events, in 2014 (*Irish Daily Mail*, April 29, 2014) and trips by government ministers to promote environmental messages on St Patrick's Day (*Irish Times*, March 10, 2007).

Coverage of the subtopic of EU-level politics falls into two broad categories: the first, in which climate change is seen as a matter for regulations and targets rather than larger societal change, and a second, in which climate change is portrayed as a supranational threat against which the EU must unite. Despite the finding that media coverage of climate change peaks around international climate conferences, there are relatively few references to such conferences in texts coded to the political or ideological contest frame. Articles dominated by this frame refer to UN COPs (eight references), G7, G8, and G20 summits (five references), or UN politics (three references) are sparsely represented in politically framed coverage in the Irish print media.

The coverage devoted to these international summits presents the issue of climate change as a metric by which the performance of political leaders may be measured. This trend of personification has been identified as a journalistic norm (Galtung and Ruge, 1965) and as a feature of climate change imagery used by the mass media (Smith and Joffe, 2009; O'Neill and Smith, 2014). In a report concerning the G7 summit in Krün, Bavaria, the relative potency of German Chancellor Angela Merkel and then French President François Hollande are assessed:

True to the Merkel method, the final G7 communiqué was an incremental agreement that made modest steps forward on pressing global issues, in particular UN agreements looming this year on climate change and development goals. First, Merkel set aside time for a public breakfast with Barack Obama, a clear effort to restore transatlantic ties burdened by the Snowden revelations. But in doing so Merkel only emphasised how she has become Obama's go-to leader in Europe. French leader François Hollande has some climate-change momentum for end-of-year talks in Paris.

(Irish Times, June 9, 2015)

In addition to representing climate change as a measure of the efficacy of particular political leaders, the issue is also "personified" as a measure of national standing in relation to the climate issue. An example of the framing of climate change as a global contest among political and economic forces occurs in a report about COP17 in Durban:

> China has raised the stakes in the game of climate poker being played here by yesterday affirming publicly and for the first time that it would be willing to sign up for a legally-binding international agreement to combat global warming. On a blustery day in the South African city of Durban, China's chief negotiator, Xie Zhenhua, told a packed press briefing his country, the world's number one carbon emitter, would be prepared to negotiate such a deal if the EU and others renewed the Kyoto protocol. Mr Xie, who is vice-chairman of China's National Development and Reform Commission, spelled out a number of other conditions, including the delivery of promised aid to poorer developing countries to help them cope with the impacts of climate change.
>
> (*Irish Times*, December 6, 2011)

In considering the subtopics dealt with in articles coded to the political or ideological contest frame, it is clear that coverage of domestic and international politics represents climate change as a policy issue offering opportunities for gaining political advantage or inflicting political damage. The coverage at EU level portrays climate change as an issue in which narrow sectoral interests must be defended against onerous regulation, while also suggesting it is a bloc-wide challenge in the face of which Europe must unite. The strategy framing is also prevalent, with a winners-and-losers narrative strongly present, and the issue is personified both by individual leaders and countries.

Special pleading: the agriculture frame

The agriculture frame was relatively sparsely represented in the corpus of 706 climate change articles. It was present in just 34 stories (4.83% of total), and was the dominant frame in 19 of these (2.69%). It was most prevalent in the *Irish Independent* (11 instances, 4.82% of frames present in that publication) and the *Irish Examiner* (8 instances, 5.93%), while the *Irish Times* (11 instances, 1.93%) and the *Irish Daily Mail* (1 instance, 2.78%) recorded low percentage representation of this frame.

Looking at the secondary frames present in the 19 stories dominated by the agriculture frame, it is noteworthy that frames emphasising policy or technical responses to climate change are strongly present (29%), as are frames emphasising opportunity (21%) and economic aspects (17%). Articles dominated by the agriculture frame are not overly concerned with politics or ideology (8%), morality or ethics (4%), or disaster (4%). An acknowledgement that climate change is anthropogenic in origin and requires urgent action is well represented in agriculture stories (13% of secondary frames), while frames undermining the scientific consensus on climate are relatively rare (4%).

It would appear that agriculture, as represented in the context of climate change, is seen as a matter largely confined to the area of policy and technocratic management. Agriculture is also constructed as a sector offering economic and other opportunities to farmers as they consider responses to the issue. A portrayal of those engaged in agriculture as being involved in the stewardship of the environment or of having responsibilities to future generations is largely absent from media coverage of the topic. The secondary framing emphasises dry, technical, or policy aspects of climate change.

A closer examination of the text contained in the agriculture frame shows that references to agriculture's contribution to Ireland's emissions occur most frequently (17 instances), followed by references to politics, both at EU and national level (13 instances). It is interesting to note that there are several references to how climate change impacts on agricultural production, but only one direct reference to agriculture's impact on the climate. The references to emissions from agriculture occur in the context of EU targets and national emissions profiles rather than in the context of climate impacts (Figure 4.5).

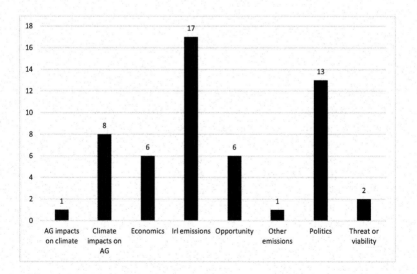

Figure 4.5 Subtopics in articles coded to agriculture frame.

Several articles make reference to GHG emissions from agriculture and to various proposals to reduce them, such as growing biomass or planting forestry to offset the GHGs produced by dairy and beef production. There are also multiple references to the argument put forward by the Irish Farmers' Association and others that reducing the size of the national herd in Ireland is of limited benefit if the demand for beef and dairy products is to be satisfied by producers in other countries who farm in a less carbon efficient way. A typical example of media representations of this argument occurs in the *Irish Times* on October 14, 2014, which quotes the then Minister for Agriculture Simon Coveney (Fine Gael party):

> "We are not talking about taking agriculture off the hook in terms of its responsibilities, but to simply force Ireland to reduce its emissions from agriculture by reducing its herd size misses the point in terms of what we're trying to achieve," he said. "This is a global problem. If we produce less, someone else somewhere else in the world will fill that gap, and in all likelihood that someone else will be producing food with a higher carbon footprint than Ireland." Mr Coveney added that Ireland had the lowest carbon footprint per litre of milk in the EU.

A concern expressed by farmers and farming organisations is that, were Ireland to reduce its beef production, meat imported from Brazil would take its place. Again, the argument is made that weaker environmental standards elsewhere would lead to an overall increase in global emissions were Ireland to attempt to cut its own emissions from agriculture. This argument is made in an article appearing in the *Irish Times* on November 26, 2007:

> The social and environmental damage associated with Brazilian beef production is totally ignored by those promoting this trade, including retailers, but it is clearly documented. The UN Commission on Human Rights has reported on the worker exploitation and slave labour problem on Brazilian ranches. Environmentalists have linked the five-fold increase in beef exports from Brazil in recent years with the rapid destruction of the rainforests in the Pantanal and Amazon regions, which is a major cause of global climate change. Brazilian beef fails to meet EU standards. It fails to meet the standards demanded and expected by European consumers. It exposes Europe to the unnecessary risk of foot and mouth disease. The evidence requires that it be banned.

These arguments constitute a subtle form of communitarianism by arguing against emissions reduction measures in Ireland on the basis that producers elsewhere will "fill the gap." Communitarian arguments usually support the advancement of national interests by placing them at a higher level of importance than global or cosmopolitan ones. By suggesting that global interests will be harmed by Ireland taking mitigation measures in the agricultural sector, this argument appears to wrap a communitarian argument in cosmopolitan clothing.

Other articles suggest that climate change will have negative impacts on Irish agriculture, such as an article published in the *Sunday Tribune* on October 21, 2007, outlining a variety of crop and animal diseases that had already been detected in Ireland. The article quotes a soft fruit farmer concerned about the milder winters caused by climate change and the effect of higher winter temperatures on his crop, and includes an analysis by Professor John Sweeney of Maynooth University that lower summer rainfall will affect potato production.

An article in the *Irish Examiner* on December 18, 2010, encapsulates the inherent contradiction in Ireland's climate policy: the impossibility of reducing emissions while increasing agricultural output. The article concerns a climate change bill the Green Party was bringing to government:

> The Green Party leader said the bill is aimed at delivering on the Government's international commitments to reduce carbon emissions. The short-term target is to reduce greenhouse gas emissions by an average of 2.5% annually by 2020 compared to 2008. The medium-term target is to reduce carbon emissions by 40% by 2030. The long-term target is an 80% reduction by 2050, compared to 1990 emissions. Mr Gormley said: "The structure of the bill provides a strong legislative framework for a core objective on transition to a low-carbon, climate-resilient and environmentally sustainable society. I am acutely aware of particular concerns in the agricultural sector, but I believe the bill poses absolutely no threat to the sustainable future of agriculture in Ireland."

Responses to the proposed legislation are then sought from farming and environmental groups. John Bryan, then the Irish Farmers' Association president, points out that the aims of the bill run counter to the growth strategy for the agricultural sector contained in the Harvest 2020 policy document, and states that any emissions reduction measures initiated in the agricultural sector would damage the economy and lead to job losses:

> This proposed legislation flies in the face of the Government's own expansion plans for the agriculture sector as set out in the Food

Harvest 2020 Report and will have a hugely damaging effect on the recovery of the economy. The proposals fail to recognise the many positives around agriculture, especially our sustainable model of farming and the carbon sink in both our permanent pasture and our forestry. Mr Bryan said that the climate change legislation fails to include a proper calculation of greenhouse gas emissions, as it fails to include the positive impact of Ireland's grassland base and forestry in its environmental impact calculations. John Bryan said: "It is ludicrous that Ireland could introduce emission reduction targets way in excess of those proposed by other countries, while at the same time countries such as Brazil destroy Amazonian rainforests and allow their greenhouse gas emissions spiral out of control."

In conclusion, it is evident that agricultural framings of climate change are dominated by technical measurements of emissions and by national political considerations. Climate impacts on agricultural production feature strongly in the coverage, while the impacts on the climate from agricultural emissions are not explicitly described. There is an implicit acceptance that such emissions are harmful, but the discourse quickly moves on to questions about the legitimacy of the emissions targets themselves or the wisdom of attempting to implement them. The rhetorical strategy of the IFA in suggesting that low carbon intensity agriculture in Ireland would be replaced by higher carbon intensity production in Brazil is also strongly present in the coverage. It is also apparent that potential beneficial impacts of climate change are presented, yet the potential benefits of a low-carbon or decarbonised agricultural sector are not.

Sceptics and sarcasm: the contested science frame

The contested science frame has a significant presence in the Irish media's coverage of climate change. There are 94 instances of this frame in the corpus of 706 articles mentioning climate change, accounting for 5.72% of all frames. A total of 78 articles (11.04% of the corpus) contain this frame. It is the dominant frame in 47 articles, accounting for 6.65% of dominant frames. So, although this frame is not strongly present in Irish coverage, it is nonetheless significant and dominates a larger number of stories than other, more prevalent frames such as the policy or technical frame (184 instances, but dominant in 45 articles) or the opportunity frame (111 instances, dominant in 30 articles).

It has been argued that the presence in the media of sceptic perspectives on the existence of climate change, its anthropogenic element, or the wisdom of mitigation or adaptation measures contributes to public uncertainty about the issue, leading to a lack of individual engagement and political action (Moser and Dilling, 2004; Painter, 2015). Therefore, media attention to contestations of climate science, or to the wisdom of climate action, may have wider societal impacts.

In the secondary framing of climate change in articles dominated by the contested science frame, the economic frame is strongly represented. In one-quarter of stories dominated by contested science, the economic frame appears as a secondary frame. Settled science is the next most prevalent secondary frame, present in 15% of cases. It is helpful to note in this context that the contested science frame contains reference to sceptic views, including occasions on which the author mentioned such views in order to dismiss or counter them. As Nisbet remarks: "...frames as general organising devices should not be confused with specific policy positions; any frame can include pro, anti, and neutral arguments, though one position might be more commonly used than others" (2009, p. 18).

The 94 instances of this frame were also analysed to discover the subtopics referred to in the relevant media texts. Some 11 subtopics were identified:

i **Confusing weather and climate**: references, sometimes humorously intended, contrasting cold weather with "global warming";

ii **"Crackpot" environmentalists**: references to environmentalists, Green campaigners, politicians, or so-called "eco-warriors" as fringe elements, easily dismissed because they are on the periphery of mainstream discourse;

iii **Discrediting climate scientists**: references to corruption among climate scientists or suggestions that scientists falsify findings for financial reward or are reluctant to speak out for fear of losing funding;

iv **Disputing climate science**: references which suggest that there is significant disagreement among scientists regarding anthropogenic global warming or which attribute climate change to solar activity or other causes;

v **Distraction from more pressing concerns**: suggestions that climate change is merely one of many global issues demanding attention, such as population growth, energy, or food;

vi **Honest broker**: this subtopic refers to a journalistic persona in which the writer assumes the mantle of an "Everyman" to ask seemingly innocent and well-meaning questions about the validity of climate science;

vii   ***Hubris***: references to the impotence of mankind in the face of the grandeur of nature, implying that our species is incapable of agency regarding the climate;

viii ***Mitigation unwise or counterproductive***: references to various mitigation measures, such as the adoption of renewable energy sources, as ineffective or possibly dangerous;

ix   ***Positive impacts***: references to climate change as a benefit to people or territories;

x   ***Religion or orthodoxy***: references to a belief in the reality of anthropogenic global warming and the necessity for urgent action as a dogma whose adherents are intolerant of diverse views; and

xi   ***Reportage on sceptic views or organisations***: reports on the statements or actions of sceptics which do not express support for such views or which mention these views in order to refute them.

Media texts contained in the contested science frame and which communicate direct or indirect contestations of climate science present a broad range of established sceptic arguments. In Rahmstorf's (2004) typology of climate denial, they are mostly trend sceptics (who deny climate change is occurring) or attribution sceptics (who accept it is happening but dispute the anthropogenic element). Often, the edifice of climate science is dismissed in its entirety as an impenetrable "ism" or "ology."

Two particular subtopics dominate contested representations of climate change: disputing climate science (30 instances) and reports concerning the activities or statements of sceptics (26 instances). Subtopics which dismiss concern over climate change as being either a fringe concern or part of a quasi-religious orthodoxy – without, it should be noted, dealing with the substantive findings of climate science – are also strongly present in the coverage, accounting for a total of 28 instances.

Several articles coded to the contested science frame suggest that climate change is not occurring or that it is a result of solar activity. For instance, an article by columnist Kevin Myers in the *Irish Independent* on May 6, 2008, states that weather data confirms the non-occurrence of global warming: "...what we hear of weather around the world does suggest that the simple slide to disaster so beloved of Warmism's doom-merchants is simply not occurring." This sentence contains two separate aspects of sceptic discourse: that climate change is not happening and that climate science is an ideology (an "ism").

Myers returns to the topic of climate change several times during the period under study, often to dismiss concerns over climate impacts as unwarranted. Myers does not deal with specific findings of climate scientists, but rather offers broad characterisations of these fields of scientific inquiry as contradictory and alarmist. For instance, he ridicules the number of scientific disciplines involved in climate science by adding "kitchensinkology" to oceanography and others, and states that calls for adaptation or mitigation measures amount to "worldwide hysteria" (*Irish Independent*, January 4, 2011).

A more explicit reference to the theory that solar activity is causing increased temperatures on Earth occurs in a discursive comment article on the agreement between the Fianna Fáil party and the Green Party to enter government together following the 2007 general election in Ireland. The author (political reporter Jody Corcoran) makes sarcastic reference to Fianna Fáil's environmental record in government before concluding that emissions-reduction measures sought by the Green Party may be unnecessary:

> The harder truth is that we are now powerless, even if we accept the argument that the emission of greenhouse gasses is mainly responsible for global warming. In fact, the Earth is getting hotter because the sun is burning more brightly than at any time during the past 1,000 years. A study by Swiss and German scientists suggests that increasing radiation from the sun is responsible for recent climate changes. Most scientists agree that greenhouse gases from fossil fuels have contributed to the warming of the planet in the past few decades but have questioned whether a brighter sun is also responsible for rising temperatures. It is doubtful that even Bertie Ahern can convince the sun to shine less forcefully. And for all their great idealism, it is doubtful that the Greens will make a significant impact on the other, man-made, causes of climate warming.
>
> (*Sunday Independent*, June 17, 2007)

There are several instances of coverage which accepts the climate is changing, but treats the suggestion that human activity has anything to do with it as an open question. Such references are often parenthetical, as in the following example from the "An Irishman's Diary" column in the letters page of the *Irish Times*:

> The latest suggestion is that it also has 400 years of fossil fuel resources, at a time when the world is wondering what part – if any – man plays in global warming and what part is down to the natural and recurring phenomenon of climate change, and the spinning of the Earth on its axis.
>
> (December 21, 2009)

Several articles could be described as presenting an "honest broker" approach, in which the author presents themselves as an "Everyman" willing to be convinced by "either side" in a hypothetical argument about the existence of anthropogenic global warming or as an honest guide to the reader trying to understand scientific complexities. These arguments often begin with an exposition of the scientific case for climate change, before entering several sceptic caveats. Such an approach characterised an article on climate change in which the author begins by presenting themselves as a disinterested citizen attempting to arrive at a sensible conclusion:

> I still can't decide whether the environmentalist lobby is a type of pseudo-religious complex or a genuine scientific inquiry. End-of-the world-itis has manifested itself with regularity down the ages. The current global obsession with it - especially in the developed world - may be some sort of spiritual manifestation in the face of the overwhelming reality of global secular materialism. Like most religious believers, environmentalists seem riddled with guilt. One can't argue with thermometers: the globe is warming as it has done before in the past, but the question now is: why? The CO_2 argument has become almost unstoppable. Despite many significant scientific sceptics, it has already generated a significant global industry.
>
> (*Sunday Business Post*, December 20, 2009)

A similar representation of the author as a typical "Everyman" occurs in an article concerning the efficacy of individual efforts to reduce GHG emissions. In this case, the author – a feature writer and music critic – adopts a passive persona of one awaiting persuasion rather than that of an active citizen prepared to seek information:

> I'm one of what I suspect to be a growing number of people who are uneasy with the subject. I'm under no illusions that the human race has done huge damage to the environment but nobody is able to offer me empirical evidence as to the effectiveness of trying to reduce one's carbon footprint. And what about carbon offsetting, another of those bandied about terms? How effective is that. Is it all a load of codswallop, a PR initiative that promises much but delivers little?
>
> (John Meagher, *Irish Independent*, May 10, 2008)

Yet another example of the "honest broker" approach to presenting sceptic representations of climate change occurs in an article written by William Reville, an associate professor of biochemistry and science awareness officer at University College Cork and author of a long-running column on science in the *Irish Times*. Reville concedes his lack of expertise in climate science yet states the importance of entertaining the views of sceptic scientists. The consensus among climate scientists that human activity is the main cause of observed climate change since the 1950s is 97% (Cook et al., 2013), and many of the remaining 3% have been shown to have ties to the fossil fuel industry (McCright and Dunlap, 2014; Oreskes and Conway, 2010).

> There is a scientific consensus that the world is gradually warming and the majority of scientists believe that this warming is caused by man-made emissions of greenhouse gases, principally carbon dioxide. However, a not-insignificant minority of scientists disagree and believe that the current warming phase is caused largely by other factors, for example changes in solar activity. I am not expert enough in meteorology and climatology to critically adjudicate on this matter and, so, I accept the conclusions of the majority of experts in this area. However, I strongly believe that we must listen thoughtfully to the minority of independent experts who advance contrary evidence.
>
> (*Irish Times*, January 15, 2009)

There are 13 instances of text presenting weather and climate as interchangeable. There has been some debate over the use of the terms "climate change" and "global warming" in communicating about changes to Earth's climate. For instance, Frank Luntz, a political strategist, suggested to the Republican Party in the US that "global warming" sounded more threatening and that Republican politicians should use "climate change" instead. A member of a focus group run by Luntz remarked that "climate change 'sounds like you're going from Pittsburgh to Fort Lauderdale.' While global warming has catastrophic connotations attached to it, climate change suggests a more controllable and less emotional challenge" (Luntz, 2002, p. 142). Of course, the terms are not interchangeable: "global warming" refers to the rise in the average mean temperatures on the Earth's surface, while "climate change" refers to a range of climactic and environmental changes caused by the trapping of GHGs in the Earth's atmosphere. One of these changes is the rise in global temperatures. Thus, global warming is a consequence of climate change: "Global warming refers to surface temperature increases, while climate change includes global warming and everything else that increasing greenhouse gas amounts will affect" (National Aeronautics and Space Administration, 2008; Wayne, 2015). However, using the terms has different effects on audiences (Ding et al., 2011; Villar and Krosnick, 2011), and presents the possibility to rhetorically undermine the existence of climate change during spells of cold weather.

Several examples of the rhetorical use of unseasonable weather to delegitimise climate science occurs in several columns by Kevin Myers: "It was the coldest April for more than 20 years: so what is the meaning of this thing, 'global warming'?" (*Irish Independent*, May 6, 2008); "We have become the Falklands of the North Atlantic, yet we apparently are still experiencing 'global warming.'" (*Irish Independent*, March 26, 2013). Myers is a controversial figure, having left the *Irish Times*, where he had written the "Irishman's Diary" column for many years, to join the *Irish Independent*. Most of the Myers articles in this corpus date from his time at the *Independent*, but more recently he moved again, this time to the Irish edition of the *Sunday Times*. On July 30, 2017, he was sacked following the publication of a column offensive to Jewish people (Logue and Gallagher, 2017). In general, Myers employs a two-fold approach when writing about climate change: in contrasting cold conditions in the present with long-term warming trends, he confuses weather and climate, and he makes scornful references to environmental policies such as energy efficient light bulbs or renewable energy (*Irish Independent*, January 8, 2009).

Often, the references to cold weather as a refutation of climate science are made as fleeting asides. An example of such parenthetical dismissal of climate change is found in a fashion article from the *Irish Daily Mail* on April 8, 2013: "In a climate such as ours – and I think we can all agree, as we shiver into April, that we can forget global warming – tweed is the most natural, and attractive, defence against the elements." Another such example occurs in a column by Maurice Neligan. Neligan, since deceased, was a retired cardiac surgeon with a prominent media profile. He was the author of a weekly column in the *Irish Times* on medical matters and the concerns of bourgeois life, such as wine and travel. He wrote in the *Irish Times* on February 17, 2009: "It snowed in Dublin last night. I wondered if anybody had told God about global warming. If so, I wish He would get His act together because we're freezing down here."

There is also a strain of commentary in journalistic content coded to the uncertain science frame which seeks to dismiss those concerned about climate change as unworthy of serious consideration. These references often refer to "lunatic" environmentalists or "nutters." They comprise a blanket delegitimisation, a characterisation of the writer as being among a sane majority, and those with whom they disagree as being "other." For instance, an article weighing the advantages of reducing one's carbon footprint by taking fewer flights, quoted Ryanair chief executive Michael O'Leary as "denouncing 'environmental nutters' who were 'persecuting' the aviation industry" (*Irish Independent*, January 13, 2007). An article on COP15 in Copenhagen contains a characterisation of climate activists as "unspeakably superior moral-high-ground environmentalists who dwell on their 'infallible' climatic knowledge" of whom the author is "weary" (*Sunday Business Post*, December 20, 2009).

A typical example of this representation occurs in a business-page article about an attempt by the Ryanair airline to distance itself from its image as aggressive towards its customers. The article contains a quote from the company's chief executive Michael O'Leary on the subject of citizens concerned about the environment:

> We want to annoy the fuckers whenever we can. The best thing you can do with environmentalists is shoot them. These headbangers want to make air travel the preserve of the rich. They are luddites marching us back to the 18th century. If preserving the environment means stopping poor people flying so the rich can fly, then screw it.
>
> (*Irish Independent*, April 2, 2014)

In a sarcastic aside, Maurice Neligan remarks that "Mother Ireland is rearing them yet" in relation to objections by environmentalists to involvement by a leading biologist in a project in the Burren (*Irish Times*, December 23, 2008). He also makes sarcastic reference to "Minister Gormley...cycling to Poland to get global warming back on track," a project he described as "Bloody nonsense" (*Irish Times*, December 16, 2008). The same author considers climate mitigation measures in a subsequent article in which he looks forward to a holiday:

> To make it worse some humourless bore will start prating to us about global warming and saving the planet and explaining to us how some penal carbon tax will make us all feel better. Such lectures had better avoid our holiday centres as anybody coming out with that doleful rubbish would run a high risk of being run out of town on a rail. As it is we'll have to wait for the next election for that satisfaction.
>
> (*Irish Times*, August 4, 2009)

A related strand of contested science content presents climate science as an orthodoxy or a religion whose adherents are not amenable to argument. Sometimes climate science is compared to Marxism or communism, such as this reference to noted climate sceptic Bjorn Lomborg in the *Irish Independent* on January 13, 2007: "Bjorn Lomborg claimed in his seminal book *The Sceptical Environmentalist* that environmentalism is an ideology just like Marxism, which also pretended to be a 'scientific' theory of economic relations." Kevin Myers also compares climate science to Marxism as well as to Catholicism, suggesting supporters of climate action are immune to sceptic arguments (*Irish Independent*, May 6, 2008). Maurice Neligan also compares supporters of climate action to religious extremists in his *Irish Times* column of December 23, 2008: "Easy on the Yule logs Sir Walter, in this year of grace such behaviour would bring some jihadist out of the woodwork screaming about global warming." Columnist Mary Ellen Synon, writing in the *Irish Daily Mail* on March 11, 2013, repeats a frequently used sceptic tactic of pointing to the so-called Little Ice Age (roughly 1350–1850) as an example of dramatic climate change occurring without human interference:

> Innocent VIII is, however, mocked today for his 1484 Bull against witchcraft. But consider his position. Climate change was leading to crop failure and mass starvation. The conventional wisdom of the time blamed the temperature changes on witches. In fact, we now know that this period between about 1300 and the late 1400s, known as the Little Ice Age, was likely to have been caused by low sun-spot activity. But at the time, witches and their incantations destroying crops was sold as the explanation. Just as now industrialists and their fossil fuel are sold as the explanation. It is likely Innocent was no more misguided in 1484 than the anti-capitalist witch-hunters of the Green movement are now.

Other texts coded to the contested science frame contain characterisations of climate change suggesting that climate change will in fact bring beneficial effects to Ireland; that the media, political, and civil society focus on climate change is a deliberate tactic to distract attention from other issues; that climate scientists are dishonest or corrupt; and that mankind is guilty of hubris in thinking that the human species is powerful enough to affect nature. In the following section, examples of these characterisations are briefly set forth:

i *Positive impacts*: There are three examples of this characterisation of climate change. One (*Irish Independent*, July 31, 2010) states that the Blasket Islands were created by a post-glacial rise in sea levels and that "if rising sea levels can produce such a captivating scene, the effects of global warming can't be all bad." The two others concern the *Irish Daily Mail*'s interview with Professor Richard Tol (November 2, 2013), analysed earlier. Tol suggests that increased levels of CO_2 will be beneficial for agricultural output;

ii **Distraction from more pressing concerns**: This characterisation of climate change occurs three times in the corpus of 47 articles containing a contested science frame. An article in the *Irish Times* on May 25, 2008, is typical of this approach, suggesting that many in the developing world are too concerned with day-to-day survival "to spend precious time worrying about ecological niceties." Kevin Myers argues (*Irish Independent*, June 24, 2009) that climate change is such a preoccupation of the UN because the greater problem of overpopulation is too contentious to be discussed;

iii **Discrediting climate scientists**: There are six instances of texts suggesting that climate scientists are corrupt or conspiratorial. Three of these (*Sunday Business Post*, December 6, 2009; *Irish Times*, February 4, 2010; *Irish Independent*, July 31, 2010) refer to the "Climategate" controversy as evidence of climate scientists' dishonesty. A letter to the editor of the *Irish Independent* (December 31, 2007) states that "the only people as irretrievably corrupt as politicians are scientists," and therefore the writer has been sceptical about "the current global warming PR blitz"; and

iv **Hubris**: There is a single instance of a hubristic representation of climate change. In his column in the *Irish Independent* on January 4, 2011, Myers contrasts a heavy snowfall with "the worldwide hysteria over global warming," before concluding that "We are nothing before the great forces of the world" and musing on "the humiliating insignificance of man."

It is evident that, in the text coded to the contested science frame, the works of Kevin Myers, and to a lesser extent, those of Maurice Neligan, are prominent in putting forward sceptic views about the reality and impacts of climate change. The work of John Gibbons is also prominent in attempting to expose and refute sceptic arguments. The tendency to dismiss climate science completely by means of sarcastic asides or ad hominem characterisations of those concerned about the environment is also evident. The range of sceptic positions identified by Rahmstorf – trend, attribution, and impact – are present in Irish media coverage of climate change, but so too is a considerable body of journalistic work challenging these positions.

References

Ahern, B., Harney, M. and Sargent, T. (2007) *Programme for Government, 2007–2012*. Dublin: Government of Ireland. Available at https://www.taoiseach. gov.ie/attached_files/Pdf%20files/Eng%20Prog%20for%20Gov.pdf

Bowe, B.J. et al. (2012) 'Framing of climate change in newspaper coverage of the East Anglia e-mail scandal', *Public Understanding of Science*, 23(2), pp. 157–169.

Cook, J. et al. (2013) 'Quantifying the consensus on anthropogenic global warming in the scientific literature', *Environmental Research Letters*, 8(2), p. 024024.

Cullinane, M. and Watson, C. (2014) 'Irish Public Service Broadcasting and the Climate Change Challenge', *Report for RTE Audience Council*. Dublin, (February), pp. 1–30.

Culloty, E. et al. (in press) 'CCIM: Climate Change in the Irish Media'. *Report for Ireland's Environmental Protection Agency*.

Dahl, T. (2015) 'Contested science in the media: Linguistic traces of news writers 'Framing Activity', *Written Communication*, 32(1), pp. 39–65.

DARA Climate Vulnerability Monitor (2012) Available at https://daraint.org/ climate-vulnerability-monitor/

Denham, B.E. (2014) 'Intermedia attribute agenda setting in the New York Times: The case of animal abuse in U.S. horse racing', *Journalism & Mass Communication Quarterly*, 91(1), pp. 17–37.

Devitt, C. and Neill, E.O. (2017) 'The framing of two major flood episodes in the Irish print news media: Implications for societal adaptation to living with flood risk', *Public Understanding of Science*, 26(7), pp. 872–888.

Ding, D. et al. (2011) 'Support for climate policy and societal action are linked to perceptions about scientific agreement', *Nature Climate Change*, 1(9), pp. 462–466.

Environmental Protection Agency (2013) *Addressing Climate Change Challenges in Ireland*. Report. Available at http://www.epa.ie/pubs/reports/ research/climate/Addressing%20Climate%20Change%20Challenges%20 in%20Ireland.pdf

Fox, E. and Rau, H. (2016) 'Climate change communication in Ireland', in Nisbet, M. (ed.) *Oxford Encyclopedia of Climate Change Communication*, pp. 1–32.

Galtung, J. and Ruge, M.H. (1965) 'The structure of foreign news. The presentation of the Congo, Cuba and Cyprus crises in four Norwegian Newspapers', *Journal of Peace Research*, 2(1), pp. 64–91.

Harcup, T. and O'Neill, D. (2001) 'What is news? Galtung and Ruge revisited', *Journalism Studies*, 2(2), pp. 261–280.

Iyengar, S. (1991) *Is Anyone Responsible? How Television Frames Political Issues*. Chicago, IL; London: University of Chicago Press.

Logue, P. and Gallagher, C. (2017) 'Sunday Times drops Kevin Myers and apologises for offensive article', *Irish Times*, 30 July.

Luntz, F. (2002) 'The environment: A cleaner, safer, healthier America'. *Report by The Luntz Research Companies*. Alexandria, pp. 131–146. Available at https://www.motherjones.com/files/LuntzResearch_environment.pdf

McAllister, L. et al. (2017) *World Newspaper Coverage of Climate Change or Global Warming, 2004–2017, Center for Science and Technology Policy Research, Cooperative Institute for Research in Environmental Sciences*, University of Colorado. Available at http://sciencepolicy.colorado.edu/icecaps/research/media_coverage/world/index.html

McCright, A.M. and Dunlap, R.E. (2014) 'Challenging global warming as a social problem: An analysis of the conservative movement's counter-claims state', *Social Problems*, 47(4), pp. 499–522.

McNally, B. (2015) 'Media and carbon literacy: Shaping opportunities for cognitive engagement with low carbon transition in Irish media, 2000–2013', *Razon y Palabra*, September(91), pp. 2000–2013.

Minihan, M. (2015) *A Deal with the Devil: The Green Party in Government.* Dublin: Maverick House.

Moser, S.C. and Dilling, L. (2004) 'Making climate hot: Communicating the urgency and challenge of global climate change', *Environment, Science and Policy for Sustainable Development*, 46(10), pp. 32–46.

Mullally, G. et al. (2013) 'Fear and loading in the Anthropocene: Narratives of transition and transformation', in *Sustainability in Society conference, University College Cork*.

National Aeronautics and Space Administration (2008) *What's in a Name? Global Warming vs. Climate Change, NASA website.*

Nisbet, M.C. (2009) 'Communicating climate change: Why frames matter for public engagement', *Environment: Science and Policy for Sustainable Development*. Taylor & Francis, 51(2), pp. 12–23.

Nisbet, M.C. (2010) 'Knowledge into action: Framing the debates over climate change and poverty', in D'Angelo, P. and Kuypers, J.A. (eds.) *Doing News Framing Analysis: Empirical and Theoretical Perspectives*. New York; London: Routledge, pp. 43–46.

Nisbet, M.C. and Scheufele, D.A. (2009) 'What's next for science communication? promising directions and lingering distractions', *American Journal of Botany*, 96(10), pp. 1767–1778.

Nuccitelli, D. (2014) 'Climate contrarians accidentally confirm the 97% global warming consensus', *Guardian.co.uk*, 5 June. Available at www.theguardian.com/environment/climate-consensus-97-per-cent/2014/jun/05/contrarians-accidentally-confirm-global-warming-consensus

O'Neill, S. and Smith, N. (2014) 'Climate change and visual imagery', *Wiley Interdisciplinary Reviews: Climate Change*, 5(1), pp. 73–87.

O'Neill, S. et al. (2015) 'Dominant frames in legacy and social media coverage of the IPCC Fifth Assessment Report – supplementary material', *Nature Climate Change*, 2, pp. 1–9.

Oreskes, N. and Conway, E.M. (2010) *Merchants of Doubt*. London: Bloomsbury Publishing.

Painter, J. (2015) *Media representations of uncertainty about climate change*. Doctoral dissertation, University of Westminster.

Rahmstorf, S. (2004) 'The climate sceptics', Report by the *Potsdam Institute for Climate Impact Research* for Munich Re, pp. 76–82. Available at http://

www.pik-potsdam.de/~stefan/Publications/Other/rahmstorf_climate_sceptics_2004.pdf

Robbins, D. (2016) 'Morality versus Politics: The Irish Media and Laudato Si'', *Studies: An Irish Quarterly Review*, 105(420), pp. 441–450.

RTE (2014) 'RTE News announces George Lee as New Agriculture and Environment correspondent', *RTE Press Centre*, 13 March.

Schäfer, M. S. et al. (2016) *Investigating Mediated Climate Change Communication: A Best-Practice Guide*. Report. Jönköping University School of Education and Communication. Available at http://www.zora.uzh.ch/id/eprint/131746/1/Schafer_et_al._-_Investigating_Mediated_Climate_Change_Communication.pdf

Schmidt, A., Ivanova, A. and Schäfer, M.S. (2013) 'Media attention for climate change around the world: A comparative analysis of newspaper coverage in 27 countries', *Global Environmental Change*, 23(5), pp. 1233–1248.

Semetko, H.A. and Valkenburg, P.M. (2000) 'Framing European politics: A content analysis of press and television news', *Journal of Communication*, 50(2), pp. 93–109.

Smith, N.W. and Joffe, H. (2009) 'Climate change in the British press: The role of the visual', *Journal of Risk Research*, 12(5), pp. 647–663.

Sönke, K. et al. (2015) *Global climate risk index 2016: Who suffers most from Extreme weather events? Weather-related loss events in 2014 and 1995 to 2014*. Briefing paper. Bonn: Germanwatch. Available at https://germanwatch.org/fr/download/13503.pdf.

Tankard, J.W.J. (2001) 'The empirical approach to the study of media framing', in Reese, S.D., Gandy, O.H. and Grant, A.E. (eds.) *Framing Public Life*. Mahwah, NJ; London: Lawrence Erlbaum, pp. 95–106.

Villar, A. and Krosnick, J.A. (2011) 'Global warming vs. climate change, taxes vs. prices: Does word choice matter?', *Climatic Change*, 105(1), pp. 1–12.

Wagner, P. and Payne, D. (2015) 'Trends, frames and discourse networks: Analysing the coverage of climate change in Irish newspapers', *Irish Journal of Sociology*, 2, pp. 1–24.

Wayne, G.P. (2015) 'Global warming vs climate change', *Skeptical Science website*, 1 July. Available at www.skepticalscience.com/climate-change-global-warming.htm

5 Ministers, handlers, and hacks

The competition to frame climate change

Introduction

I have argued elsewhere that the perspectives of journalists are often neglected or absent from research into media coverage of news topics. Researchers are focused on the published text and, in the case of framing studies, on the frames contained therein. However, before journalistic texts are composed and published, other complex processes are at work. Various strategic actors attempt to influence editors and reporters and to persuade them to adopt particular issue frames. Other constraints, norms, and cultures influence the process of assembling news content. In the following sections, interviews with journalists (both environmental correspondents and newspaper editors) are presented with a view to exposing some of these processes to establish a dominant framing of climate change in the media.

Interviews are often used to deepen analysis and expose different perspectives (Morse, 1991; Thurmond, 2001; Olsen, 2004), and interviews with journalists are also a common form of data collection, although they are often presented as the main or only data rather than a means of triangulation of other, more quantitative information (see, among others, Fahy, O'Brien, and Poti, 2010 and Pihl-Thingvad, 2015 for examples of interview-based research). Some studies of media coverage of climate change have included interviews with journalists covering the issue, but these were concerned with scientific literacy and accuracy (Wilson, 2000; Boykoff and Mansfield, 2008), did not foreground the views of journalist interviewees (Mormont and Dasnoy, 1995; Aykut, Comby and Guillemot, 2012), or concentrated on journalists in the developing world and their personal attitudes to climate change (Harbinson, Mugara and Chawla, 2006). It is evident from the corpus of research on media coverage of climate change that there has been a concentration on the replication of scientific accuracy, and perspectives from science communication and environmental communication have been privileged. There have been several calls for research which includes the perspective of journalists covering climate change (for instance, Anderson, 2009, p. 176; Dahl, 2015, p. 60).

The interview subjects for this thesis were chosen with an eye to scholarship examining the role of the media in the development of complex social problems and, in particular, those theories which address the notion of issue competition. Several scholars have suggested that various actors attempt to dominate the framing of particular issues in ways that favour their agenda, and thereby gain "issue ownership" (Gusfield, 1981) and prevail in a battle for the scarce resource of public and policy attention (Hilgartner and Bosk, 1988; Baumgartner and Jones, 1991). The media, politicians, NGOs, and policymakers are among those engaged in this competition, with the media often the target of the framing strategies of the other groups. Thus, the media play a central role in these issue competitions, and the development of issue coverage in the media from niche specialist reporting to broader, more politicised coverage is a key factor in raising public awareness (Nisbet and Huge, 2006). Journalists who covered the environmental "beat" during the time frame under study were considered important interview subjects, as were those engaged in the issue competition concerning climate change from outside the media, such as politicians active on environmental issues and political advisors dealing with the media. The perspective of newspaper editors overseeing coverage of climate change was also considered valuable. The list of interview subjects therefore comprised the environment correspondents of influential media organisations such as the state broadcasting organisation RTÉ (Paul Cunningham), the *Irish Times* (Frank McDonald), *Irish Independent* (Paul Melia), and *Irish Examiner* (Claire O'Sullivan), the editors of the *Irish Times* (Kevin O'Sullivan) and *Sunday Independent* (Cormac Bourke), and a prominent columnist and blogger on climate (John Gibbons). O'Sullivan's views were of interest as he was employed as environment correspondent prior to his appointment as editor of the paper and has returned to the same "beat" since his resignation as editor in April 2017, albeit with the new title of environment editor. Thus, he has experience both as a reporter trying to "pitch" stories to an editor and as an editor on the other side of this process. Bourke's perspective is also valuable, as he worked as news editor on the *Irish Independent* before he was appointed editor of the *Sunday Independent*. He was designated to speak on behalf of the Independent group of newspapers by the then editor-in-chief Stephen Rae. Paul

Cunningham, now a news anchor on RTÉ radio's flagship news programme *Morning Ireland* and author of a book on the impacts of climate change on Ireland (Cunningham, 2008), was also a key interview subject. Non-media actors interviewed comprised two former government ministers (John Gormley, former minister for the Environment, Heritage and Local Government and Eamon Ryan, former minister for Energy, Communications and Natural Resources, both serving in office from 2007 to 2011) who were in office for much of the time frame examined by this thesis, and their departmental media advisors (Liam Reid, John Downing, and Bríd McGrath). The perspectives of these actors allow for access to privileged information and expert insight from those most closely involved in climate change communication in recent times.

Climate change as a difficult journalistic topic

Some events and issues are more attractive to the media because they display certain characteristics that align them with journalists' notions of what "makes a good story." These "news values" include discreet events (as opposed to long-term trends) that happen close by to people who resemble the story's intended audience (Galtung and Ruge, 1965; Harcup and O'Neill, 2001). Drama and the presence of heroes and/or villains (Jones, 2014) add to the likelihood of inclusion on the news agenda. These news values have approximate equivalents in the scholarship of framing. For instance, the news values of immediacy and discretion are echoed in the preference for episodic rather than thematic frames; thus, the framing choices of journalists bear out the news values of their profession. An example of this preference for coverage to focus on single, discreet events rather than placing them in longer-term contexts occurred with the reporting of two extreme weather events in the Caribbean and Gulf of Mexico in autumn 2017 (hurricanes Harvey and Irma), whereby coverage focused on the individual hurricanes and largely missed the opportunity to place them in the larger content of climate change (Radtke, 2017).

There was widespread agreement among interview subjects that, as a journalistic topic, climate change did not align with news values adhered to in newsrooms. For instance, climate change is perceived as invisible, distant in time and often geographically; it is complex and data-heavy; it is "for nerds, by nerds, to nerds" (Bríd McGrath, personal interview, February 8, 2017). The lack of a direct, visible impact of climate change in Ireland was also seen as a barrier to media coverage, which in turn rendered the issue imperceptible and esoteric to an Irish media audience. According to one interviewee: "It was niche, people didn't really believe it and they thought it might affect Mauritius, the Seychelles, Bangladesh and small island nations somewhere far away but not us" (Ibid.).

The long timelines involved in climate change also made it difficult for journalists to cover: "It's very difficult to write about something, a front-page story about something that's going to happen potentially in about 20 years' time" (Claire O'Sullivan, personal interview, June 16, 2015). This view was echoed by a veteran environmental journalist:

> ...you're talking about the future rather than the present as such, even though there're elements of the future in the present, in the sense of the typhoons in the Philippines and all that kind of stuff, that may or may not be attributable to the climate change, but it's the kind of stuff that's likely to happen with more regularity as time goes on. So yes, it is a difficult thing to cover because a lot of people are focused on short-term issues and that includes politicians. And politicians do not have a long-term view in general.
> (Frank McDonald, personal interview, May 15, 2015)

Apart from the imperceptibility of climate change, in Ireland at least, and the long time horizons predicted for some impacts, the repetitious nature of climate change as a media topic can be a factor in editorial decisions on coverage. John Gormley recalled speaking at the first COP (in Berlin in 1995) and saying that

> we have 10 years to do something about this and then, of course, at Copenhagen we were saying we've only 10 years left... The messages are so doom-ey and gloomy, so people are saying, 'Well, you said that 10 years ago and we're still around, so, you know, what's the story here?'
> (John Gormley, personal interview, April 13, 2015)

It can seem, as several interview subjects stated, that climate change is the same story over and over again with slight changes to climate data or small increments to scientific modelling projections:

> ...if you're only having confirmation by ways of scientific evidence then there is a sense from an editor's [point of view]: 'Haven't we seen that before, haven't we done that before?' To take it in TV terms, in 2005 we went out to Greenland, and we went to the

Jakobshavn glacier, which calves the tallest vertical icebergs in the world, and we were out there with the icebergs and we were talking about the melt, but if I was to go back now, I'm sure, or if George [Lee, currently RTÉ's agriculture and environment correspondent] tried to do it now, there would be a sense of 'We've done that already, the Arctic ice, we did that, we spent the money on it.' So, there's a repetitious nature to some of the evidence and that militates against editors who are always looking for something new. It doesn't seem to be new.

(Paul Cunningham, personal interview, April 24, 2017)

The repetitious nature of the story of climate change as a homogenous entity during the early part of the time frame under consideration led to difficulties, especially for broadcast journalists such as Paul Cunningham. He spoke about the dearth of visual "hooks": once you had broadcast footage of polar bears and melting Arctic ice, there were few alternatives: "How many times can I use the Polar ice cap? You can't, so you have to find something new" (Paul Cunningham, personal interview, April 24, 2015). It also posed a challenge for columnists writing about environmental matters on a weekly basis: "…over that period I was essentially trying to find ways of presenting the same argument week after week and trying to find different ways to engage people, and what might work and so on" (John Gibbons, personal interview, May 1, 2015). The news values inherent in journalistic practice affected the nature of climate change coverage. Although IPCC reports and climate data were published at intervals, in essence the "story" of climate change remained the same, even though details and particular data might change: GHG emissions were causing the global temperature to rise. Furthermore, a relatively small cohort of sources was available to journalists, and visual imagery associated with the topic was becoming hackneyed (Paul Cunningham, personal interview, April 24, 2015). The prospect of pitching a climate change story to an editor given these constraints was summed up by one environmental correspondent:

… because if you went up to an editor and said, 'listen, I've got a climate change story' you know they're going, in their mind, 'this is going to have John Sweeney [IPCC author and geographer, Maynooth University], it's going to have a fucking polar bear, it's going to have some Arctic ice, and it's going to have Paul saying "we're all going to die."'

(Ibid.)

This aspect also has implications for public engagement with the issue. Studies on consumer behaviour have shown that continued exposure to the same message may provoke a negative reaction (Tellis, 1997; Campbell and Keller, 2003), while extended media coverage of the same topic can lead to the exhaustion of new sources and information (Vasterman, 2005). Specialist environmental journalists may even become disenchanted with a news subject over extended periods (Djerf Pierre, 1996). Taken together, these factors (public resistance to oft-repeated message, scarcity of "new angles" for journalists, and disenchantment among journalists) may combine to create "issue fatigue" (Djerf-Pierre, 2012, p. 501).

The news value of proximity – that events taking place close to the site of publication are selected above those happening in distant locations – also influenced climate change coverage. One environmental reporter mentioned that she would have encountered resistance to suggestions for coverage of distant environmental impacts had she suggested them. She stated that she "would have been laughed out of it" if she had suggested doing a story on the impacts of climate change on the Arctic, for instance. Even more local stories, such as flooding risks in Cork's docklands area, were declined:

> That is it again: if anything is esoteric, if anything is long-term, they [editors] have no interest. But once people can feel it, and it's practically proven, then you could [pitch stories] ...When it was just esoteric, it was more difficult. But once it started being translated into: well you might not be able to afford to run two cars, you mightn't be able to afford to work 30 miles away from home, when it was actually impacting on people's lives and their aspirations, then I think it was getting more inches in the paper, you know.
>
> (Claire O'Sullivan, personal interview, June 16, 2015)

It was apparent in the interviews that journalists were advocating for the environment in certain situations, especially when pitching stories to their superiors. On occasion during the interviews, the issue of climate coverage and climate action was conflated by the interview subject, giving the impression that they intended coverage as a motivating factor towards action. This throws up concerns of journalistic balance and objectivity.

Although journalism which is perceived to be biased is the source of much criticism of the media (Blumler and Cushion, 2014), it may be more valued by journalists themselves, who strive to be impartial in order to remain credible (Deuze, 2005), than it is by the public (Gil de Zúñiga and Hinsley, 2013). There is abundant evidence that newsrooms create their own culture, and newspapers and media organisations transmit unspoken attitudes to social issues to their reporters (Breed, 1955; Tuchman, 1972; Reese and Ballinger, 2001), making objectivity impossible in any case.

The journalists interviewed for this research fulfilled an advocacy role in so far as they wished to see levels of coverage increased so that the public could be more informed about the issue of climate change. In some cases, they commented negatively on the readiness of their news editors to accept climate-related news stories. Most were aware of the danger of "balance as bias" (Boykoff and Boykoff, 2004) in reporting debates about climate science, as if scientific opinion were divided equally on the issue, instead of 97% versus 3%. Indeed, most stated that they never sought such "balance" in their reporting by contacting sources who denied climate science or advocated inaction on climate change. They also exemplified the role of specialist correspondent identified by scholars as those who are "treated more as independent experts, free to make judgments" (Schudson, 2001, p. 163), or who are seen as star reporters, independent and removed from the "vast journalistic sub-proletariat" of ordinary reporters (Bourdieu, 1998), and who have the ability to shape coverage "away from asking basic questions about the science to exposing the political games and pressures" (Carvalho and Burgess, 2005, p. 1465).

Journalists are reputed to be poor at mathematics and their professional culture has prized language skills over numeracy (Harrison, 2016), even though reporters have the knowledge but lack the confidence to deal well with data-heavy topics (Maier, 2003). Journalists' perceived lack of numeracy makes a topic such as climate change inherently unattractive. As a ministerial advisor put it: "It's data heavy, it is big, there's hockey sticks, hockey curves, you have various people out who are saying there's nothing happening" (Liam Reid, personal interview, May 19, 2015). One interviewee, who has high numeracy skills, remarked that he is now tasked with reporting on other data-heavy topics, such is the dearth of numerate newsroom colleagues (Paul Melia, personal interview, August 6, 2017). Thus, the statistical complexity of climate change, added to the other aspects of the topic which do not align with journalists' news values, make it a challenging topic for journalists to write about.

How journalists frame climate change

It is useful to bear in mind that theorists differ on whether framing is an unconscious activity (Lakoff, 2009) or entails some deliberate elements of selection and exclusion (de Blasio and Sorice, 2013). When it comes to journalists, there may be elements of both dynamics at play: journalists, through frequent repetition as part of their work routines, may utilise certain frames semi-automatically on some stories, while on other unusual or complex stories they may have to exercise conscious judgment. Some of the journalists interviewed for this thesis tended to adopt framing strategies common in their news organisations. For instance, newspapers such as the *Irish Independent* and *Sunday Independent* had a "house style" when it came to framing most issues. Their reporters used the economic frame, with a special emphasis on financial impacts on individuals or households (Paul Melia, personal interview, August 6, 2017; Cormac Bourke, personal interview, August 1, 2017). Others were more deliberative in their framing approaches, trying out creative approaches in order to satisfy their editors (Paul Cunningham, personal interview, April 24, 2015; Claire O'Sullivan, personal interview, June 16, 2015).

A senior editor also confirmed that a so-called "punter-friendly" approach was common in his media group, suggesting it was almost a template for coverage of many issues (Cormac Bourke, personal interview, August 1, 2017). An example cited by one interviewee, which falls outside the scope of this thesis, was the publication of Ireland's National Mitigation Plan (Government of Ireland, 2017) in July 2017. The document was criticised by the Climate Change Advisory Council for a lack of policy proposals (Climate Change Advisory Council, 2017). In an example of their policy of relating complex, abstract issues to the everyday lives of their readers, the *Irish Independent* covered the story by highlighting a proposal to reduce the speed limit on motorways as a means of reducing CO_2 emissions from the transport sector (Paul Melia, personal interview, August 6, 2017).

Opportunity and political framings could also attract the attention of editors. Coverage which emphasised the potential of responses to climate change to create jobs was also put forward by the environment correspondent of the *Irish Independent*, who suggested that stories of entrepreneurial success in the environmental sector were likely to be published. Writing about the politics of climate change was also a means by which the topic could be presented to readers. On occasion, this coverage could relate to Ireland's emissions targets:

> there was also a political side to it, there was the question of European targets and Ireland's inability to match those targets and so were able to bash ministers up, saying, 'You were going to do X, you were going to do Y.'
>
> (Paul Cunningham, personal interview, April 24, 2015)

The annual Conferences of the Parties (COPs) also provided an opportunity to present preview articles looking forward to the event, coverage from the event itself, and reflective pieces in the event's aftermath. The conferences also provided journalists with access to national politicians. For one journalist, they "were a focus. We also had delegates going to them, so 'Who were they?' 'What were they saying?' 'What aspects were they going to be reporting on?'" (Ibid.). For another, however, they failed to provide an opportunity for coverage: "COPs? No. Just even the word COP, no. Absolutely not" (Claire O'Sullivan, personal interview, June 16, 2015).

Journalists still working in the area stated that climate change could be related to other topics, such as extreme weather, planning, energy and water policy, and waste and resource management. One journalist cited the example of proposals to reduce the number of diesel-powered vehicles in Ireland, which is put forward primarily as a health concern.

> ...the push on diesel is more about air quality than climate, but obviously, there's both in it. And we need to kind of, I think, look at more the climate change debate in terms of just an environmental and health debate. Not necessarily that it's climate change, it's doom and gloom for future generations. Climate change includes quality of life for all. Having less cars on the streets of Dublin is good for people in Dublin. It's a nicer place to walk around, it's easier to get around. If we're talking about farming, you know, being less polluting in farming means there's less sewage and the rivers, it's cleaner, there's more wildlife, there's more biodiversity. I think we might need to start framing it like that.
>
> (Paul Melia, personal interview, August 6, 2017)

A newspaper editor suggested that dealing with environmental issues was a means of bringing climate change to public attention without addressing it directly. The management of resources such as energy and water, the introduction of energy-saving measures in the built environment, changes to motor tax regimes, and similar topics can be used to present aspects of climate change:

> ...you can talk to people about the big, bad wolf of climate change and they don't quite know what it means...the best way to tell people about the bigger picture is to give them something they can understand from their own end. I think that's the general principle, certainly for papers like the *Irish Independent* and the *Sunday Independent*.
>
> (Cormac Bourke, personal interview, 2017)

One environment correspondent makes an effort to present stories about planning, energy, and flooding in the context of climate change (Paul Melia, personal interview, 2017).

There is a disparity between how journalists describe their own framing strategies and the findings relating to the presence of frames in media texts. First, reporters stress the economic framing of the issue, generally suggesting that successful coverage addresses financial impacts on individuals and family finances. However, the data shows that political frames dominate the coverage of climate change in the Irish print media. This disconnect between the nature of the actual coverage and journalists' perceptions about how they frame it suggests that the political framing may take place at the subconscious level, or at least be semi-automatic, the "default frame" so to speak, while other frames, such as the opportunity or economic frames, require conscious decision-making on the part of the reporter. Second, it is evident that the primary audience as far as the journalist is concerned is not the reading public but the commissioning editor to whom they report. Their framing strategies are aimed first and foremost at this managerial level rather than at a wider audience. The requirement to satisfy fellow journalists as a first step in the publication of climate change coverage serves to reinforce journalistic norms and news values and maintain a journalistic culture of coverage (Schultz, 2007; Gade, 2008). A third aspect of journalistic perspectives relates to the depoliticisation and re-politicisation debate referred to in the previous chapter (Chapter 5, section 5.19). Several journalists suggest that climate change may be covered without making the story "about" climate change (i.e. by focusing on constituent issues such as waste, water, public health, or flooding). This is akin to the "third way" of communicating about climate change, threading a path between polarising political debates and outright denial (Nisbet, 2013).

Yes, Minister: framing strategies from within the cabinet

The framing strategies of politicians and political operatives were investigated by means of semi-structured interviews with two Green Party ministers, their respective media advisors, and the former deputy government press secretary. The two Green Party ministers interviewed had differing approaches to communicating about climate change. The differences were apparent both from their own responses to interview questions and from the perspectives of their media advisors. John Gormley, the former Minister for the Environment, Heritage and Local Government, sought to emphasise actions that individuals could take to mitigate climate change. Among the initiatives his department introduced were a transition to an emission-based motor tax regime, changes to energy-efficient light bulbs, and a carbon tax on fossil fuels. He also initiated a public information campaign linking climate change to individual lifestyle choices, and suggested that the Green Party ministers choose hybrid vehicles over conventional cars as their ministerial transport. Given this foregrounding of individual emissions reduction actions, it was predictable that the media would seek to expose any hypocrisy in Gormley's personal behaviour or in that of the other Green Party representatives. The *Irish Daily Mail* in particular engaged in a campaign of undermining the legitimacy of the Green Party's position on climate change by seeking to highlight any party activity that caused excessive (or indeed any) GHG emissions. Furthermore, the employment by Gormley of two former journalists as media advisors may have been counterproductive, as it led to an abrasive relationship between the media and the minister and his staff. A further complicating factor in assessing Gormley's communications strategy is his role as party leader, which required him to engage with the media on party issues such as ministerial and Seanad appointments, the Greens' relationship in government with Fianna Fáil, and its reaction to various pronouncements by Fianna Fáil Taoisigh (party leaders) Bertie Ahern and Brian Cowen. The media's coverage of Gormley, therefore, is based on more than his performance and qualities as a minister.

Eamon Ryan, the former Minister for Communications, Energy and Natural resources, emphasised narratives of ecological modernisation in his communications with the media. Each departmental communication contained a mention of climate change, but technological solutions, rather than individual actions, were foregrounded. Ryan's activities in office concerned expansion of the energy grid, the introduction of new postal codes, the establishment of investment incentives for renewable energy, and the introduction of electric vehicles. These were presented in a positive, business-friendly context.

Similarly, the media advisors to both ministers had different approaches. John Downing, the deputy government press secretary, found that he was not called upon to deal with media queries concerning climate change; the bulk of his dealing with the media concerned politics. Liam Reid, who worked more closely with Gormley on media relations, was required to head off or otherwise deal with negative coverage of the party leader. Whatever framing strategies they may have favoured were secondary to "reacting, firefighting, trying to manage, trying to dial down, trying to deal" (John Downing, personal interview, February 23, 2017).

Bríd McGrath, who worked with Ryan, had more success with a framing of economic opportunity in her communications with the media. She also employed a tactic of strategic placement of stories, allocating events, announcements, and policy proposals to a series of specialist publications and specialist correspondents. She also stated that not having worked as a journalist was an advantage, as journalists treated her more professionally and her relationship with them was less hostile than those of her colleagues.

In summary, Gormley and his team sought to communicate climate change as a real danger the solution to which lay partly in changing individual behaviour. The emphasis on personal behaviour invited media scrutiny of Gormley's own behaviour. Ryan and his advisor stressed the economic and opportunity frames, which was a more successful media strategy as it avoided the stereotype of environmentalists "talking down" to their audience or lecturing them from on high.

The crowding-out effect and the decline in media interest

In considering the dramatic decline in the coverage of climate change in Irish media from December 2009, the "crowding-out" effect identified by Monica Djerf-Pierre (whereby environmental topics are crowded out by financial news and reports of conflict) is a more plausible explanation than Downs's attention cycle model. However, the evidence is not clear-cut. The financial crisis began to impact on Ireland from late 2008, and indeed coverage in the UK and Europe suffered a steep decline in November 2008. Levels of coverage in Ireland dropped, but not dramatically so, and reached their highest level during the time frame under study in December 2009. However, from late December 2009, Irish coverage declined severely.

Many of the interviewees pointed to the economic crash of late 2008–2009 as the key factor for the decline in media coverage of climate change at that time. One correspondent recalled the impact of the banking collapse and the arrival of the troika of the European Commission, the European Central Bank, and the International Monetary Fund in Ireland on the area of environmental journalism:

> ...suddenly I couldn't get a story on to save my life. The very idea of being concerned with something that could happen in 100 years was just being eclipsed completely by thousands of people have been thrown out onto streets today, people losing their homes today, it was today's story, so talk about momentum towards deals, anything like that was seen as completely irrelevant in the context

of the times. And I was trying...I was struck...I remember at about the end of 2008, there was about three months where I was nearly considering saying 'Do you want me do business stories?' because I was offering stories which had regularly got on air and I just couldn't get them on. It didn't matter: TV, radio, online. No one gave a monkey's.

(Paul Cunningham, personal interview, April 24, 2015)

This view is shared by those actively involved in trying to attract media coverage for the issue, such as media advisors.

I don't blame the media for not noticing climate change when they thought the world was collapsing around us. Everyone at a particular time maybe from, want to say, early 2009 on were buying the FT, opening the back of it and looking up bond yields. That's what was going on, not carbon things, not how hot is the world today, but bond yields...If you turned up at a particular time and start going on about climate change, eyes rolled, 'Shut up, stop saying the same thing again, look at us, we're about to get cut adrift into the ocean or we won't be able to pay the civil servants tomorrow' or whatever. So there became a time when it annoyed people to talk about climate change...We were like a Cormac McCarthy novel, like it was holocaust territory...I think it was just a sad, insular, depressing time where we thought we lost the sovereignty of the nation and yes, that trumped climate change.

(Bríd McGrath, personal interview, February 8, 2017)

From the perspective of news values and journalistic norms, the financial crash in Ireland was an all-consuming story. It had many of the attributes required to make it attractive to journalists: drama, novelty, meaningfulness, and reference to something negative (Galtung and Ruge, 1965; Harcup and O'Neill, 2001). One government minister closely involved in efforts to deal with the financial crisis recalled the intense media focus on the economic story:

At that point, we were in such feckin' chaos in terms of [laughs] they [the media] didn't give a shit about any climate bill. The only story in town was Troika, IMF, bailout of banks, whatever, bondholders being burnt, whatever. For pretty much from 2008 onwards, October 2008 onwards, there wasn't any other story in Ireland.

(Eamon Ryan, personal interview, February 1, 2017)

The decline in climate change coverage does not map neatly onto the timeline of the financial crash in Ireland. For instance, the first major event in the credit crisis was the issuing of a guarantee to cover the liabilities of seven banks operating in the jurisdiction in September 2008, yet coverage of climate change remained relatively steady afterwards. Yes, there is a decrease from June 2008, but it was incremental. In December 2009, in the run-up to the Copenhagen COP, coverage reached its highest point. It could be argued that the decline post-December 2009 is as much related to the deflation of politicians and journalists at the disappointing outcome in Copenhagen as to the effects of the ongoing financial crisis.

How the "frame competition" around climate change plays out

The competitive framing environment around climate change in the time frame of this study was a complex one. The framing strategies of the media, government ministers, and their advisors were discussed earlier in isolation; in this section, how they performed when coming into contact with one another is discussed. Gormley attempted to frame the issue as one in which individual lifestyle choices were a key means of tackling rising emissions. This is similar to the responsibility frame identified in studies of generic frames (Semetko and Valkenburg, 2000). In a significant contribution during our interview, Gormley, speaking about a public information campaign, said that "...we were trying to make them see, to make the connection between what they were doing in their daily lives and the big, big issue of climate change" (John Gormley, personal interview, April 13, 2015). Ryan, on the other hand, emphasised the economic and opportunity frames. Gormley's media advisors were forced into reactive strategies by the responsibility framing adopted by their minister, while Ryan's advisor was able to be more proactive and strategic in her framing.

The media, meanwhile, reported on climate change largely as a political or ideological contest. They engaged in what could be called meta-coverage: they did not, to any great degree, adopt the frames put forward by politicians or media advisors, but reported on the two competing framing strategies through their own preferred game or strategy frame. Thus, the substantive issues of the impact of personal actions or of the economic impacts or opportunities of climate change were not engaged with; rather, a meta-narrative of how these tactics were succeeding relative to one another in the political arena was presented. The media did not undertake journalism which considered the problem and investigated the relative merits of proposed solutions; instead they covered only the politics of climate policy, essentially writing about the contest between Gormley and Ryan interpretations of the issue – meta-coverage that bypassed the topic itself.

References

Anderson, A. (2009) 'Media, politics and climate change: Towards a new research agenda', *Sociology Compass*, 2(3), pp. 166–182.

Aykut, S.C., Comby, J.-B. and Guillemot, H. (2012) 'Climate change controversies in French mass media 1990–2010', *Journalism Studies*, 13(2), pp. 157–174.

Baumgartner, F.R. and Jones, B.D. (1991) 'Agenda dynamics and policy subsystems', *The Journal of Politics*, 53(4), pp. 1044–1074.

Bell, M. (1998) 'The journalism of attachment', in Kieran, M. (ed.) *Media Ethics*. London: Routledge, pp. 15–22.

de Blasio, E. and Sorice, M. (2013) 'The framing of climate change in Italian politics and its impact on public opinion', *International Journal of Media and Cultural Politics*, 9(1), pp. 59–69.

Blumler, J.G. and Cushion, S. (2014) 'Normative perspectives on journalism studies: Stock-taking and future directions', *Journalism: Theory, Practice & Criticism*, 15(3), pp. 259–272.

Bourdieu, P. (1998) *On Television and Journalism*. London: Pluto Press.

Boykoff, M.T. and Boykoff, J.M. (2004) 'Balance as bias: Global warming and the US prestige press', *Global Environmental Change*, 14(2), pp. 125–136.

Boykoff, M.T. and Mansfield, M. (2008) '"Ye Olde Hot Aire": Reporting on human contributions to climate change in the UK tabloid press', *Environmental Research Letters*, 3, p. 024002.

Breed, W. (1955) 'Social control in the newsroom: A functional analysis', *Social Forces*, 33(17), pp. 326–335.

Campbell, M.C. and Keller, K.L. (2003) 'Brand familiarity and advertising repetition effects', *Journal of Consumer Research*, 30(2), pp. 292–304.

Carvalho, A. and Burgess, J. (2005) 'Cultural circuits of climate change in UK broadsheet newspapers, 1985–2003', *Risk Analysis*, 25(6), pp. 1457–1469.

Climate Change Advisory Council (2017) 'Letter to Minister Denis Naughten', pp. 1–4. Available at www.climatecouncil.ie/media/Letter%20to%20 Minister%20Naughten%20March%202017.pdf

Cunningham, P. (2008) *Ireland's Burning: How Climate Change Will Affect You*. Dublin: Poolbeg Press.

Dahl, T. (2015) 'Contested science in the media: Linguistic traces of news writers "framing activity"', *Written Communication*, 32(1), pp. 39–65.

Deuze, M. (2005) 'Popular journalism and professional ideology: Tabloid reporters and editors speak out', *Media, Culture & Society*, 27(6), pp. 861–882.

Djerf Pierre, M. (1996) 'Green news: Environmental reporting in television 1961–1994', *Unpublished thesis*. Department of Journalism and Mass

Communication, University of Gothenburg. Available at https://jmg.gu.se/publicerat/bokserie/09_grona_nyheter

Djerf-Pierre, M. (2012) 'The crowding-out effect: Issue dynamics and attention to environmental issues in television news reporting over 30 years', *Journalism Studies*, 13(4), pp. 499–516.

Fahy, D., O'Brien, M. and Poti, V. (2010) 'From boom to bust: A post-Celtic Tiger analysis of the norms, values and roles of Irish financial journalists', *Irish Communications Review*, 12(1), pp. 5–20.

Gade, P.J. (2008) 'Journalism guardians in a time of great change: Newspaper editors' perceived influence in integrated news organizations', *Journalism & Mass Communication Quarterly*, 85(2), pp. 371–392.

Galtung, J. and Ruge, M.H. (1965) 'The structure of foreign news. The presentation of the Congo, Cuba and Cyprus crises in four Norwegian Newspapers', *Journal of Peace Research*, 2(1), pp. 64–91.

Gil de Zúñiga, H. and Hinsley, A. (2013) 'The press versus the public: What is "good journalism?"', *Journalism Studies*, 14(6), pp. 926–942.

Government of Ireland (2017) *National Mitigation Plan*. Available at https://www.dccae.gov.ie/en-ie/climate-action/publications/Documents/7/National%20Mitigation%20Plan%202017.pdf

Gusfield, J. (1981) *The Culture of Public Problems: Drinking-Driving and the Symbolic Order*. Chicago, IL: University of Chicago Press.

Harbinson, R., Mugara, R. and Chawla, A. (2006) 'Whatever the weather: Media attitudes to reporting climate change', *Report*. Panos Institute, London, p. 20. Available at http://panoslondon.panosnetwork.org/wp-content/files/2011/03/whatever_weathermjwnSt.pdf

Harcup, T. and O'Neill, D. (2001) 'What is news? Galtung and Ruge revisited', *Journalism Studies*, 2(2), pp. 261–280.

Harrison, S. (2016) 'Journalists, numeracy and cultural capital', *Numeracy*, 9(2). Available at http://scholarcommons.usf.edu/numeracy/vol9/iss2/art3/

Hermida, A. and Domingo, D. (2011) 'The active recipient: Participatory journalism through the lens of the Dewey-Lippmann debate', conference proceeding, *International Symposium on Online Journalism*, University of Texas, Austin, Texas.

Hiles, S.S. and Hinnant, A. (2014) 'Climate change in the newsroom: Journalists' evolving standards of objectivity when covering global warming', *Science Communication*, 36(4), pp. 428–453.

Hilgartner, S. and Bosk, C.L. (1988) 'The rise and fall of social problems: A public arenas model', *The American Journal of Sociology*, 94(1), pp. 53–78.

Jones, M.D. (2014) 'Cultural characters and climate change: How heroes shape our perception of climate science', *Social Science Quarterly*, 95(1), pp. 1–39.

Lakoff, G. (2009) *Why Environmental Understanding, or 'Framing,' Matters: An Evaluation of the EcoAmerica Summary Report*, Huffingtonpost.com.

Maier, S.R. (2003) 'Numeracy in the newsroom: A case study of mathematical competence and confidence', *Journalism & Mass Communication Quarterly*, 80(4), pp. 921–936.

Mormont, M. and Dasnoy, C. (1995) 'Source strategies and the mediatization of climate-change', *Media, Culture & Society*. Sage Publications, 17(1), pp. 49–64.

Morse, J. (1991) 'Approaches to qualitative-quantitative methodological triangulation', *Nursing Research*, 40(1), pp. 120–123.

Nisbet, M.C. (2013) 'Environmental advocacy in the Obama years', in Vig, N. and Kraft, M. (eds.) *Environmental Policy: New Directions for the Twenty-First Century*. 9th edn. Washington, D.C.: Congressional Quarterly, pp. 58–78.

Nisbet, M.C. and Huge, M. (2006) 'Attention cycles and frames in the plant biotechnology debate: Managing power participation through the press/policy connection', *The Harvard International Journal of Press/Politics*, 11(2), pp. 3–40.

Olsen, W. (2004) 'Triangulation in social research: Qualitative and quantitative methods can really be mixed', *Developments in Sociology*, 20, pp. 1–30.

Pihl-Thingvad, S. (2015) 'Professional ideals and daily practice in journalism', *Journalism*, 16(3), pp. 392–411.

Radtke, D. (2017) 'Sunday shows largely fail to mention climate change in Hurricane Irma coverage', *Media Matters for America*, 10 September.

Reese, S.D. and Ballinger, J. (2001) 'The roots of a sociology of news: Remembering Mr. Gates and social control in the newsroom', *Journalism & Mass Communication Quarterly*, 78(4), pp. 641–658.

Schudson, M. (2001) 'The objectivity norm in American journalism', *Journalism*, 2(2), pp. 149–170.

Schultz, I. (2007) 'The journalistic gut feeling', *Journalism Practice*, 1(2), pp. 190–207.

Semetko, H.A. and Valkenburg, P.M. (2000) 'Framing European politics: A content analysis of press and television news', *Journal of Communication*, 50(2), pp. 93–109.

Tellis, G.J. (1997) 'Effective frequency: One exposure or three factors?', *Journal of Advertising Research*, 37(4), pp. 75–80.

Thurmond, V.A. (2001) 'The point of triangulation', *Journal of Nursing Scholarship*, 33(3), pp. 253–258.

Tuchman, G. (1972) 'Objectivity as strategic ritual: An examination of newsmen's notions of objectivity', *American Journal of Sociology*, 77(4), p. 660.

Vasterman, P.L.M. (2005) 'Media-hype: Self-reinforcing news waves, journalistic standards and the construction of social problems', *European Journal of Communication*, 20(4), pp. 508–530.

Wilson, K.M. (2000) 'Drought, debate, and uncertainty: Measuring reporters' knowledge and ignorance about climate change', *Public Understanding of Science*, 9(1), pp. 1–13.

6 Communicating climate change in the new media environment

Introduction

Climate change is obviously a complex problem. It is difficult for pol-icymakers to confront, and it is also challenging for journalists to cover. Since climate change became a media topic in the summer of 1988, it has become a divisive and contested battleground, a social and political problem rather than simply a scientific one. In Ireland, there is broad acceptance in the media and among media sources of the reality of climate change – that it is happening. Ireland's official re-sponse has been to adopt the language of international climate action in public while in private arguing for less ambitious targets and special arrangements for the country.

In this volume, we have seen how Ireland's media covers climate change. The coverage is low by European standards, is politically framed, and does not engage with the deeper and more systemic im-plications of a changing climate. Are the Irish media performing their normative roles when it comes to climate change? Are they holding power to account, informing the public or sounding warnings of dan-gers ahead? I would argue that, within the constraints of their pro-fession, they are. The prominence of the settled science frame in the coverage shows that the public has been exposed to coverage which accepts the reality of anthropogenic global warming and the necessity for emissions reduction. The interviews presented here show a moti-vated and professional cohort of environmental journalists who have shown considerable ingenuity and determination in pitching stories and devising angles for their news editors. However, they are work-ing within an industry experiencing severe financial challenges, and are working in an area which does not fit well with the values of their profession.

Covering climate change: lessons from and for journalists

It is evident that media coverage of climate change is influenced to a great degree by news values. Climate change does not display many of the characteristics journalists look for in events or topics that influence their place on the news agenda. All of the journalist interviewees testified that climate change was an inherently difficult topic to cover. It is not immediate, its effects are not visible yet in Ireland, it is complex and data-heavy, and journalists see themselves as possessing poor numeracy skills. Perhaps most importantly, it is repetitive, with a repeating narrative which changes only in relation to details of data.

A further influence on the coverage is the necessity for journalists to consider the attitude of their editors (rather than their readers) as a primary concern when framing coverage. This dynamic contains climate change within a journalistic culture, and newsroom norms may take precedence over professional or even societal ones. Furthermore, journalists need "hooks" on which to hang their coverage. In other words, they need something to report on. Elsewhere, climate change-related legislation or policy announcements provide such hooks. Ireland, however, is seen as a "laggard" (Torney and Little, 2017) when it comes to policy responses to climate change, and these opportunities for coverage do not often arise. Consequently, coverage hooks become restricted to international conferences and the release of scientific reports. These news hooks in turn promote certain media frames: conferences invite political framings, while IPCC reports lend themselves readily to disaster frames (Painter, 2013, 2014).

Journalists themselves offered a timeline of coverage as they experienced it in their workplace: a general lack of interest on the part of editors in the early 2000s followed by a surge of interest from 2006 onwards. The Stern Report (2006), the release of the movie *An Inconvenient Truth* (2006), and the release of the Fourth Assessment Report of the IPCC (2007) provided raw material and pretext for coverage of the issue. My research again emphasises the need for "news hooks" on which to hang coverage and also links to the news value of "continuity" whereby a topic already in the news is more likely to receive coverage than a completely novel topic. In 2006 (the Stern Report, *An Inconvenient Truth*) and 2007 (AR4, the Green Party entering government and attending the Bali COP), climate change was in the news, and news organisations were more disposed to cover it. However, according to one environment correspondent, "by 2008, it was gone" (Paul Cunningham, personal interview, April 24, 2015).

In the absence of news hooks and seeking to gain approval from their editors for story proposals, journalists often framed their articles in ways which they knew would be congruent with the general approach of their news organisation. At the Independent group, which sees itself as representing "middle Ireland" (Cormac Bourke, personal interview, August 1, 2017), economic frames were favoured, while at the *Irish Daily Mail*, contrarian tactics were often employed. This replication of newsroom culture, taking place at the level of individual news editor or reporter, in turn served to amplify the ideological stance of the wider news organisation and led to media-wide coverage of the issue that emphasised approaches and solutions from within the economic and political status quo.

Nonetheless, journalists remain motivated to provide climate change coverage to their news organisations and showed considerable ingenuity in overcoming institutional apathy towards the subject. A frequently used strategy was to connect climate change to coverage of other, related topics such as air pollution, water quality, waste management, flooding, and planning. This disaggregation of climate change had the effect of depoliticising it, of reducing it to constituent issues about which there is little or no disagreement. The academy is divided on whether this tactic is necessary to build broad coalitions across political divides or whether such political sanitisation of climate change in fact discourages deep engagement by the public.

The divergent approaches of the media advisors to the two Green Party ministers in government from 2007 to 2011 also influenced the media's coverage of climate change. In the case of the Minister for Communications, Energy and Natural Resources, the media message emphasised the necessity of dealing with climate change through a pro-business, ecomodern approach, and coverage of this minister's policy initiatives, public statements, and appearances was positive (Bríd McGrath, personal interview, February 8, 2017). The Minister for the Environment, Heritage and Local Government foregrounded the personal lifestyle choices of the Green Party members of parliament, and his policies involved more negative measures, such as the introduction of bans and taxes. This invited the media to explore possible divergence between his public statements and his private behaviour, and resulted in more negative coverage. The employment of two former journalists as media advisors to this minister contributed to an adversarial relationship between the media and the minister's press team, fuelling further negative coverage of climate change-related policy measures.

The presence of two Green Party ministers in cabinet had a complex effect on media coverage. Journalists interviewed for this research stated that a Green voice in government provided them with more "news hooks" on which to hang coverage, and the data supports the view that, when coverage of the Green Party increased, so also did coverage of climate change. However, the Green Party "signal" in the coverage of climate change is weak, and only 9.2% of climate change coverage mentions the Green Party. Some of this coverage could be negative, and given the circumstances in which the Green Party left government, it is likely that they had a negative effect on the media's attention for climate change at that time.

The journalists, politicians, and media operatives interviewed here agree that the financial crisis which began in mid-2008 had a significant negative impact on media attention for climate change. The news values to which journalists subscribe mean that more immediate, dramatic, and negative events take precedence in the news agenda over events which do not exhibit these characteristics. However, when the views of interviewees are considered in light of the data on media attention for climate change, a more complex picture emerges. Although the effects of the financial crisis began to be felt in Ireland from the summer of 2008 (RTE News, 2008), media coverage of climate change did not decline until December 2009, in line with international trends. This suggests the "crowding-out" effect (Djerf-Pierre, 2012) may be a more drawn-out process than previously suggested. The importance of the December 2009 climate change conference in Copenhagen is likely to have been a factor in maintaining coverage levels.

There may be a further dimension to the crowding-out effect. The initial research into this aspect of media dynamics concluded that news of armed conflict and financial crises pushed environmental issues off the media agenda (Ibid.). It may be the case that attempts to communicate via the media on environmental issues at such times are counterproductive. In November 2010, the "troika" of the International Monetary Fund, the European Central Bank, and the European Commission instituted an "Economic Adjustment Programme" for Ireland, known colloquially as "the bailout." At that time, the Green Party was trying to steer a Climate Change Bill through parliament. It is evident from the statements of journalists and ministerial media advisors working at that time that not only had media attention for climate change-related matters declined, but also that attempts to increase media interest for climate change at a time of financial crisis provoked an antagonistic response. Not only were the media uninterested in climate change, they viewed attempts to promote it up the media agenda as naïve and even unpatriotic (Bríd McGrath, personal interview, February 8, 2017).

It is noteworthy that, although the data from newspaper coverage shows that political framings were by far the most common, journalists themselves suggest that they tried to present climate change through the economic frame by means of linking the impacts of climate change or of mitigation or adaptation measures to individual impacts. In fact, several interviewees downplayed the role of politics in their coverage of the issue. The two newspaper editors interviewed also suggested that climate change coverage in their particular titles was focused on impacts, often financial impacts, rather than on political considerations.

This points to a disconnect between the nature of the actual coverage of climate change and the journalists' own perception of how they and their newspaper titles present it. Environmental journalists, when asked to reflexively analyse their framing strategies, suggest that the economic frame is most successful. This is how journalists, when having to pitch stories to their editors, frame climate change. However, much climate change coverage is reactive, covering set-piece events and conferences, and thus bypasses the process whereby the reporter researches and pitches a story for inclusion in his organisation's publication. Left to their own devices, journalists favour economic framings, but when assigned to cover events, revert to political ones. As discussed earlier, the data makes it clear that significant peaks in coverage occur around UN climate conferences; these are essentially political gatherings, and the political or ideological contest frame is prominent in the coverage at these times (O'Neill et al., 2015). These events are the main drivers of climate change coverage and call forth political framings from journalists.

The need to satisfy editors as to the newsworthiness of climate change stories also means that the topic remains contained in a journalistic culture, at a remove from the scientific reality of climate change and from its social and political aspects. The institutional cultures of the Independent Group and the *Irish Examiner*, for instance, perceive climate change through the lens of impacts on individual and family incomes, while the *Irish Times* culture emphasises climate politics and a global perspective.

It is also apparent that the topic of climate change is in the process of being segmented into constituent subframes, such as planning, flooding, resource management, and waste management. These subframes are more easily aligned with traditional news values, are smaller in scale, more local in relevance, and can be linked to the larger whole of climate change by the journalist.

Alternative narratives in Ireland's climate change coverage

The journalists interviewed tended to rate or rank other media organisations in journalistic terms, assessing the coverage of various news outlets in Bourdieusian terms of capital within the journalistic field. From this viewpoint, the *Irish Times* and RTÉ (at least when Paul Cunningham was working as environment correspondent) had the most journalistic capital, while the *Irish Daily Mail*, which was seen as a subversive force, had least. Columnist and blogger John Gibbons, who combines the roles of journalist and activist, was able to place the tone and content of coverage in a wider perspective:

> The [*Daily*] *Mail* and the *Express*, they're obsessed by weather stories, by health stories. Everything gives you cancer or prevents cancer, and I think there seem to be a whole string of health and weather stories, and weather extremes, but clearly there's no interest in the underlying science whatsoever...And I think they're just blow with the wind, and I think their ownership, most of these newspapers we're talking about are owned by expat billionaires, and I do think expat billionaires have a particular issue with, how to say, global good housekeeping...I think their focus is, 'Don't pay tax and destroy regulation.' And I think when they see climate change, they hear, 'Regulation.' Like the IFA, and therefore they're fighting it, not for what it is, but for what they perceive it to be as a threat to their interests.
>
> (Personal interview, May 1, 2015)

The *Irish Daily Mail* was perceived by many interviewees as pursuing an editorial policy that sought to undermine the legitimacy of proponents of climate action. For example, the *Mail* foregrounded the expense of attending climate change conferences, and submitted freedom of information requests so that they might calculate the amount of paper consumed by Green Party activities in parliament. The press advisors to John Gormley recounted an attempt by the *Irish Daily Mail* to establish where Minister Gormley was staying for the Copenhagen COP so the newspaper could focus on the cost of rooms, suites, or meals (Liam Reid, personal interview, May 19, 2015), while the press advisor to Eamon Ryan stated that the Irish edition of the paper published reports from the UK edition concerning UK climate proposals but simply substituted the names of Green Party politicians for the names of the UK Labour Party ministers (Bríd McGrath, personal interview, February 8, 2017) appearing in the original UK edition reports. Other researchers have found that the *Mail* group is seen as untrustworthy and deliberately contrarian on this issue (Lockwood, 2009, p. 8) and seeks to portray climate change as a means of introducing "stealth" environmental taxes (Hibberd and Nguyen, 2013).

The *Irish Daily Mail* contained relatively elevated levels of the contested science and morality frames, and the texts coded to these frames contained a range of sceptic arguments and representations of climate change as a religion, suggesting that acceptance of the existence of anthropogenic global warming was a matter of faith rather than of science. Other studies have found that the *Mail* engages in a range of sceptic rhetorical repertoires, such as the "settler" (rejecting alarmism on grounds of "common sense") and comic nihilism (unserious, blithe rejection of climate science) repertoires (Ereaut and Segnit, 2006). It is clear that the *Irish Daily Mail* engaged in an editorial strategy in its climate change coverage that was markedly different from those pursued by other media organisations. Much of this coverage was presented as a campaign to expose wasteful use of taxpayers' money or to question orthodoxy on behalf of the common citizen; both these approaches also had the effect of delegitimising climate science and undermining public appetite for climate action. A close relationship between the willingness of a news organisation to publish sceptic views and its ideological alignment has been established (Carvalho, 2005, 2007; Carvalho and Burgess, 2005), although ideological orientation does not correspond neatly with levels of sceptic coverage (Painter and Gavin, 2015). The *Mail* coverage, though presented as common-sense consumer advocacy, nonetheless served the interests of its political and economic agenda.

Another alternative narrative was evident in media texts dealing with agriculture. For instance, the contribution of agriculture to Ireland's emissions profile is frequently mentioned. Agriculture contributes 33.1% of Ireland's total GHG emissions (Government of Ireland, 2017, p. 80), compared to an EU-28 average of 10% (Eurostat, 2017) and a global average of 20% (IPCC AR5). Over half of Ireland's agricultural emissions derive from enteric fermentation in ruminant animals. In the content coded to the agriculture frame, an alternative narrative about agriculture's role in Ireland's emissions emerges, with farming organisations suggesting (i) that the measurement of emissions does not adequately account for carbon sinks such as grassland and forestry found on Irish farms, (ii) proposed emissions reduction measures (such as those in legislation proposed by the Green Party in 2010) run counter to other government policy, which encourages increases in food and dairy production, and (iii) emissions reduction efforts in Irish agriculture are misguided, as any shortfall in beef or milk production in Ireland would be filled by increased production elsewhere, perhaps in territories such as Brazil, where regulations are of a lower standard and agriculture is less carbon efficient.

The narrative put forward by the Irish Farmer's Association (IFA) and the Irish Creamery and Milk Suppliers' Association (ICMSA) – that farming and the agri-food business are important to the Irish economy and therefore any measures inhibiting the growth or endangering the viability of these sections would run counter to the national interest – is similar to other economic arguments against mitigation measures. However, it does contain an implied domestication perspective, suggesting that measures adopted by Ireland alone will count for little in a global context and even that well-intentioned policies may be taken advantage of by less scrupulous rivals in the beef or dairy industry. The climate change legislation proposed by the Green Party in 2010 was singled out for particular criticism from farming organisations, as it committed Ireland to emissions reductions targets in excess of those envisaged by the European Union (*Irish Examiner*, December 18, 2010). This narrative also casts Ireland as the victim of unfair emissions measurement systems at EU level (Lynch, 2014).

The two examples just analysed – the faux concern of the *Mail* for taxpayers' money and the special pleading of the agriculture lobby – are variants of familiar media tropes. The *Mail*'s approach is a form of climate scepticism, as is the approach of the farming organisations: both are kinds of distraction from more substantive issues. In the first instance, the *Mail*'s emphasis on expenses is an attempt to portray climate change policy generation as a wasteful exercise, while ignoring the predicted impacts of climate change itself; in the second, the efforts for special concessions for Ireland's agriculture sector also fail to take wider perspectives into account. However, other opportunities for more profoundly alternative narratives have not been availed of by the media. It is clear that the opportunity frame is a salient feature of Irish coverage and offers possibilities for competing narratives about mitigation and adaptation to be presented. Yet the text coded to this frame adheres to conventional narratives suggesting that new technologies, efficiencies, and market mechanisms can be mobilised to combat climate change, thereby missing an opportunity to communicate alternative narratives to the public.

Recommendations for best practice in communicating about climate change

Some conclusions may be drawn with regard to best practice for those communicating about climate change. First, it is evident that employing a dedicated environmental correspondent serves to increase the level of coverage of climate change in a media organisation. The decline in the levels of coverage in the *Irish Times* in the absence of an environment correspondent is apparent from the data presented here, as is the increase in the levels of coverage of climate change published by the *Irish Independent* since the appointment of an active and engaged correspondent. The coverage of climate change by the national broadcaster, RTÉ, was found to decline following the departure of a dedicated environment correspondent (Cullinane and Watson, 2014).

Employing a dedicated correspondent may change the nature of a media organisation's coverage. Specialist correspondents have more autonomy in the newsroom, are freer to express opinions, and are treated as independent experts whose judgment can guide coverage (Schudson, 2001; Hiles and Hinnant, 2014). The correspondents interviewed for this book exhibited considerable resourcefulness and ingenuity in pitching stories to their editors and in deploying a range of framing strategies. They were also more active in generating original news material rather than reporting from events. Thus, specialist correspondents may help move news organisations away from an over-reliance on the strategy and conflict frames towards more positive and engaging framings, such as the opportunity and morality frames.

Second, a strong argument may be made that a reporting strategy based on coverage of the constituent elements of climate change – rather than the issue as a hegemonic whole – may be successful. Data from the interviews with journalists and editors suggest that topics such as planning, air pollution, water quality, environmental protection, energy efficiency measures, and technological advances may align more closely with the news values of editors. These topics may then be linked with the larger issue of climate change in the text of the news article.

It is apparent that journalists assigned to cover set piece events such as international climate change conferences or the release of climate reports revert to familiar political or disaster frames in their reports, whereas journalists who propose original, "off-diary" coverage have more freedom to deploy a wider range of framing strategies. Journalism which reacts to events is characterised by a more limited range of frames, while proactive, original journalism is more diverse in its framing.

For those wishing to influence media coverage of climate change, it is clear from the evidence presented here that some framing strategies are more successful than others. The responsibility frame, deployed by the former Minister for the Environment, John Gormley, proved counterproductive and produced a hostile reaction, particularly from news outlets publishing content sceptical about climate change. The economic and opportunity frames proved more successful when employed by the former Minister for Communications, Eamon Ryan. However, it proved difficult for those seeking to influence media coverage to persuade journalists to abandon their preferred conflict and strategy framings, suggesting that, while external actors may increase levels of coverage, influencing the nature of coverage may prove more difficult.

It is also evident that a flexible approach, tailoring communications material (and how such information is framed) to specific news outlets can attract coverage. The media advisor to the Minister for Communications employed such a strategy, directing differing material to television, newspapers, specialist outlets, and websites (Bríd McGrath, personal interview, February 8, 2017).

This approach was in contrast to her colleagues (John Downing and Liam Reid) working with the Minister for the Environment. These media advisors spent considerable amounts of time attempting to discourage or minimise negative coverage. Both had worked previously as journalists for print newspapers. While this experience gave them valuable insights into news routines and journalism practice, it also meant they had existing relationships with the political, environment, and news journalists they now dealt with as media advisors (John Downing, personal interview, February 23, 2017; Bríd McGrath, personal interview, February 8, 2017). This state of affairs in turn meant it was difficult to maintain an appropriate professional distance from those journalists they engaged with, and it calls into question the advisability of employing former journalists as press advisors.

Journalists covering climate change offered recommendations for improving both the level of coverage and its content. Improved scientific literacy among journalists would help in being able to summarise complex documents (Claire O'Sullivan, personal interview, June 16, 2015), and less newsroom pressure would allow reporters to investigate the issue more deeply (Kevin O'Sullivan, personal interview, August 14, 2017). Journalists should ignore sceptic arguments (Frank McDonald, personal interview, May 15, 2015), attend to the local implications of international events, and concentrate on the disparity between goals and targets and actions deployed to achieve them

(Paul Cunningham, personal interview, April 24, 2015). All, however, emphasised the difficulties in communicating the complexities of the issue.

The perspectives of media advisors in relation to media practice are also valuable. Those seeking media coverage must provide effective communication rather than "blame the conduit guys," who may be uninterested in climate change and are often "cut and paste merchants" (Bríd McGrath, personal interview, February 8, 2017). This view point links to arguments about a lack of material on which journalists can base their reports. Perhaps, instead of attempting wholesale changes in the way the media operates, it may be more productive for actors to tailor their communications to appeal to the media as they are currently constituted.

The media need something to report on. Few newsrooms, especially in Ireland, have the resources to carry out investigations ab initio. In his comparison between climate change coverage in 1988 and 2012, Sheldon Ungar pointed to a key difference between the two periods. In 1988, there were multiple "inputs" relating to climate change: events, speeches, hearings, reports, and protests (Ungar, 2014). In short, there were matters the media could report on, pretext under which climate change could be covered. In Ireland, given the lack of action on climate change, there are relatively few occasions for coverage: sparse legislation, a dearth of policy initiatives, and few parliamentary debates. If there is little action on climate change by way of reports, policy initiatives, parliamentary debates, or climate legislation, then the news hooks on which journalists can hang stories are missing. Climate change is inherently repetitive, and as a range of social theorists have noted, focusing events are required to promote issues up the agenda ladder. Journalists need these focusing events also, as a pretext under which climate change-related material can be published. The environmental journalists interviewed for this thesis are ready to take advantage of any inputs from the policy or politics spheres as a means by which climate change information can be put before the public.

Conclusion

What, in the end, can we say about Ireland's coverage of climate change? And how is Ireland's coverage relevant to other territories? In many ways, Ireland's coverage of climate change conforms to international norms. In terms of levels of media attention for climate change, Ireland's coverage exhibits the same patterns of peaks and troughs as many other countries, albeit at lower levels. In terms of coverage frames, the prominence of the political frame is also quite typical. Yet, when the coverage is examined more closely, some peculiarly Irish aspects emerge. For example, there is a national reluctance to engage with the topic at a systemic level, to really explore what is required to combat climate change, and what implications such efforts might have for society. This reluctance to engage is evident in both coverage which acknowledges the reality of anthropogenic global warming and coverage which disputes or downplays human causation. Even in sceptic coverage, the tone is one of sarcasm and dismissal rather than debate.

For a country in which agriculture occupies such a large place in the national imagination, media coverage which features an agriculture frame is surprisingly low – just 2.69% of climate stories are dominated by this frame, and it is present in just 35 stories out of 706 (4.83%). Indeed, agriculture contributes over a third of Ireland's non-ETS-traded emissions, and one might expect the coverage to reflect this state of affairs. What coverage there is tends to consider the impacts of climate change on agriculture rather than vice versa. Agriculture is "the elephant in the room" of Irish climate policy and of media coverage of climate change: a sector of the economy whose contribution to Ireland's emissions profile is increasing, but which is seldom included in policy or media discourse on the issue.

When we come to consider how Ireland's media coverage of climate change may be relevant to other territories, the data presented in this volume, both quantitative (levels of media coverage) and qualitative (frame analysis and interviews), supports the following conclusions:

i   Media coverage of climate change in Ireland is in many ways typical of a small, peripheral EU state. Ireland is a "rule-taker" from the EU, a role that applies even more obviously where climate change is concerned. Ireland's media coverage is predominantly political, but often the politics in question are at EU level, especially the politics of emissions targets and mitigation policy. Ireland expects to be told what to do about climate change by the EU, and much of its political energy has gone into reducing the level of ambition at EU level. This in turn has led to a perception that mitigation measures have been imposed on Ireland from outside, and there is little enthusiasm for implementing them, let alone for taking a leadership role on climate change. This policy agenda of reducing targets or arguing for exemptions, derogations, or other special treatment for Ireland has naturally influenced the media agenda, which has focused on EU politics to a large extent. Thus, it is likely that the media in smaller EU states will focus on political wrangling at Commission or Council of Ministers level rather than on broader environmental or societal impacts of climate change;

ii   Ireland's media system is more closely aligned to the US and UK systems than to the European journalistic tradition. Ireland, then, is part of a media system that adheres to rigid news formulations and is focused on so-called hard news, in contrast to Europe, where interpretive, discursive, and analytical journalism is more common (Chalaby, 1996). This US-UK influence contributes to newsroom cultures and journalistic work practices which are not well suited to covering complex, ongoing topics such as climate change. In terms of its size, location, and role within the EU, Ireland has much in common with the Nordic countries and some Eastern European states; in media terms, Ireland is more oriented towards the Anglophone centres of the US and the UK. This state of affairs means that Ireland is an interesting and relevant case, as it bridges the two political and media power centres of the old social democratic Europe on the one hand and the more market-oriented US-UK axis on the other;

iii   The effects of the financial crisis on Ireland's media coverage of climate change are of interest and may provide universally applicable insights. The concept of the "crowding out effect," whereby news of war and financial calamity pushes out environmental topics (Djerf-Pierre, 2012), is well established and is supported by the tracking of global news coverage of climate change (McAllister et al., 2017). Ireland's case adds some further nuance and detail to this effect. First, there may be a time lag involved in the crowding out effect. The financial crisis hit in the autumn of 2008, but the decline in media interest in climate change did not occur until the end of December 2009, suggesting that even in the face of pressing financial news, climate change can still find a place on the media agenda once there is something compelling for journalists to write about. In this case, the Copenhagen climate conference of 2009 (COP15) saw a notable peak in coverage even in the midst of a grave financial crisis in Ireland. Second, financial bad news "crowds out" climate change news (eventually), but the effect identified by Monica Djerf-Pierre may do more than that. The study of the Irish case shows that not only do the media lose interest in environmental issues at times of financial crisis, they become positively antagonistic to it. Journalists and politicians trying to push climate change coverage were seen as deluded and even unpatriotic at the time; and

iv   The effect on the media coverage of climate change of having the Irish Green Party serve in a coalition government may also be of interest to political and media scholars in countries where such a situation has occurred (much of Europe) or may happen in the future (South America and Africa). Naturally, Green Parties in government are inclined to focus on environmental issues and bring forward relevant policies to cabinet. This research shows that having the Green Party in government in Ireland did increase media attention for climate change. However, when the nature of such coverage is examined, a more nuanced picture emerges. The ways in which Ireland's two Green Party ministers communicated about climate change had differing effects on their treatment at the hands of the media. Attempts to preach or talk down to the public on individual lifestyle choices provoked negative coverage, while strategies focussing on opportunity and economics were more successful. Furthermore, although the Greens are good for climate change, at least as far as media coverage in concerned, climate change is not good for the Greens. When the Greens are in the news, so also is climate change; when climate change is in the news, it is seldom related to Green Party policy or actions.

Some aspects of Ireland's media coverage of climate change are universal: its patterns of coverage and the ways in which it is framed, for example. Others, such as its treatment of the agriculture sector and its particular brand of dismissive scepticism, are uniquely Irish. Yet others offer a focus for further study: the role of newsroom cultures on media framings, for example, or the disconnect between the ways reporters reflexively frame climate change and the ways in which they frame it in actual coverage. Up to now, Ireland has been a neglected focus of study of media coverage of climate change; I hope this volume has gone some way to rectifying this state of affairs.

References

Carvalho, A. (2005) 'Representing the politics of the greenhouse effect: Discursive strategies in the British media', *Critical Discourse Studies*, 2(1), pp. 1–29.

Carvalho, A. (2007) 'Ideological cultures and media discourses on scientific knowledge: Re-reading news on climate change', *Public Understanding of Science*, 16(2), pp. 223–243.

Carvalho, A. and Burgess, J. (2005) 'Cultural circuits of climate change in UK broadsheet newspapers, 1985–2003', *Risk Analysis*, 25(6), pp. 1457–1469.

Chalaby, J. (1996) 'Journalism as an Anglo-American invention. A comparison of the development of French and Anglo-American journalism', *European Journal of Communication*, 3, pp. 303–326.

Cullinane, M. and Watson, C. (2014) 'Irish public service broadcasting and the climate change challenge', *Report for RTE Audience Council*. Dubin, (February), pp. 1–30.

Djerf-Pierre, M. (2012) 'The crowding-out effect: Issue dynamics and attention to environmental issues in television news reporting over 30 years', *Journalism Studies*, 13(4), pp. 499–516.

Ereaut, G. and Segnit, N. (2006) 'Warm words: How are we telling the climate story and can we tell it better?', *Report for the Institute for Public Policy Research*. Available at www.ippr.org/publications/warm-wordshow-are-we-telling-the-climate-story-and-can-we-tell-it-better

Eurostat (2017) *Greenhouse Gas Emissions Statistics*. Available at http://ec.europa.eu/eurostat/web/products-datasets/-/t2020_rd300

Government of Ireland (2017) *National Mitigation Plan*.

Hibberd, M. and Nguyen, A. (2013) 'Climate change communications & young people in the Kingdom: A reception study', *International Journal of Media & Cultural Politics*, 9(1), pp. 27–46.

Hiles, S.S. and Hinnant, A. (2014) 'Climate change in the newsroom: Journalists' evolving standards of objectivity when covering global warming', *Science Communication*, 36(4), pp. 428–453.

Lockwood, A. (2009) 'Preparations for a post-Kyoto media coverage of UK climate policy', in Boyce, T. and Lewis, J. (eds.) *Climate Change and the Media*. New York: Peter Lang, pp. 186–199.

Lynch, S. (2014) 'EU climate summit to agree revision of Irish emission targets', *Irish Times*, 14 October.

McAllister, L. et al. (2017) *World Newspaper Coverage of Climate Change or Global Warming, 2004–2017, Center for Science and Technology Policy Research, Cooperative Institute for Research in Environmental Sciences*, University of Colorado, web.

O'Neill, S. et al. (2015) 'Dominant frames in legacy and social media coverage of the IPCC Fifth Assessment Report – supplementary material', *Nature Climate Change*, 2, pp. 1–9.

Painter, J. (2013) *Climate Change in the Media: Reporting Risk and Uncertainty*. Oxford: I B Tauris.

Painter, J. (2014) *Disaster Averted? Television Coverage of the 2013/14 IPCC's Climate Change Reports*.

Painter, J. and Gavin, N. T. (2015) 'Climate skepticism in British newspapers, 2007–2011', *Environmental Communication*, 10(4), pp. 37–41.

RTE News (2008) 'Lenihan admits "serious problem" in economy', *RTE News Website*, 24 June. Available at www.rte.ie/news/2008/0624/104891-economy/

Schudson, M. (2001) 'The objectivity norm in American journalism', *Journalism*, 2(2), pp. 149–170.

Torney, D. and Little, C. (2017) 'Symposium on the politics of climate change in Ireland', *Irish Political Studies*, 32(2), pp. 191–198.

Ungar, S. (2014) 'Media context and reporting opportunities on climate change: 2012 versus 1988', *Environmental Communication: A Journal of Nature and Culture*, 8(2), pp. 233–248.

Index

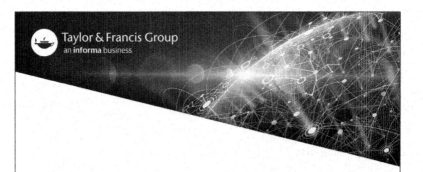